'This book is a monumental event. William MacAskill is one of the most important philosophers alive today, and this is his magnum opus.'

Rutger Bregman

'Many books promise a new "big idea", but few deliver one as brilliant as MacAskill's in *What We Owe The Future*. A fascinating, profound read.'

Julia Galef

'*What We Owe The Future* makes the case for thinking seriously about the very long term. It gives a profoundly new perspective on human civilization and our place in it.'

Lydia Cacho

'There are moments when we can change outcomes easily, but if we don't bend those curves right then, we can lock in enormous long-term damage. This fascinating book makes us think relentlessly and usefully about such pivot points; few prods could be more important.'

Bill McKibben

WHAT WE OWE THE FUTURE

Also by William MacAskill

Doing Good Better

WHAT WE OWE THE FUTURE

A MILLION-YEAR VIEW

WILLIAM MACASKILL

A Oneworld Book

First published in Great Britain, the Republic of Ireland and Australia
by Oneworld Publications, 2022

Published by arrangement with Basic Books, an imprint of Perseus Books LLC,
a subsidiary of Hachette Book Group, Inc.

A CIP record for this title is available from the British Library

ISBN 978-0-86154-250-5 (hardback)
ISBN 978-0-86154-482-0 (trade paperback)
eISBN 978-0-86154-251-2

Interior design by Amy Quinn
Printed and bound in Great Britain by Clays Ltd, Elcograf S.p.A.

Oneworld Publications
10 Bloomsbury Street
London WC1B 3SR
England

For my parents, Mair and Robin, and their parents,
Ena and Tom and Daphne and Frank, and . . .

Contents

PART I
THE LONG VIEW

Introduction

Imagine living, in order of birth, through the life of every human being who has ever lived.[1] Your first life begins about three hundred thousand years ago in Africa.[2] After living that life and dying, you travel back in time and are reincarnated as the second-ever person, born slightly later than the first. Once that second person dies, you are reincarnated as the third person, then the fourth, and so on. One hundred billion lives later,[3] you become the youngest person alive today. Your "life" consists of all of these lifetimes, lived consecutively.

Your experience of history is very different from what is depicted in most textbooks. Famous figures like Cleopatra or Napoleon account for a tiny fraction of your experience. The substance of your life is instead composed of ordinary lives, filled with everyday realities—eating, working, and socialising; laughing, worrying, and praying.

Your life lasts for almost four trillion years in total. For a tenth of that time, you're a hunter-gatherer, and for 60 percent you're an agriculturalist.[4] You spend a full 20 percent of your life raising children, a further 20 percent farming, and almost 2 percent taking part in religious rituals. For over 1 percent of your life you are afflicted with malaria or smallpox. You spend 1.5 billion years having sex and 250 million giving birth. You drink forty-four trillion cups of coffee.[5]

You experience cruelty and kindness from both sides. As a colonizer, you invade new lands; as the colonized, you suffer your lands taken from you. You feel the rage of the abuser and the pain of the abused. For about 10 percent of your life you are a slaveholder; for about the same length of time, you are enslaved.[6]

You experience, firsthand, just how unusual the modern era is. Because of dramatic population growth, a full third of your life comes after AD 1200 and a quarter after 1750. At that point, technology and society begin to change far faster than ever before. You invent steam engines, factories, and electricity. You live through revolutions in science, the most deadly wars in history,[7] and dramatic environmental destruction. Each life lasts longer, and you enjoy luxuries that you could not sample even in your past lives as kings and queens. You spend 150 years in space and one week walking on the moon. Fifteen percent of your experience is of people alive today.[8]

That's your life so far—from the birth of *Homo sapiens* until the present. But now imagine that you live all future lives, too. Your life, we hope, would be just beginning. Even if humanity lasts only as long as the typical mammalian species (one million years), and even if the world population falls to a tenth of its current size, 99.5 percent of your life would still be ahead of you.[9] On the scale of a typical human life, you in the present would be just five months old. And if humanity survived longer than a typical mammalian species—for the hundreds of millions of years remaining until the earth is no longer habitable, or the tens of trillions remaining until the last stars burn out—your four trillion years of life would be like the first blinking seconds out of the womb.[10] The future is big.

If you knew you were going to live all these future lives, what would you hope we do in the present? How much carbon dioxide would you want us to emit into the atmosphere? How much would you want us to invest in research and education? How careful would you want us to be with new technologies that could destroy or permanently derail your future? How much attention would you want us to give to the impact of today's actions on the long term?

I present this thought experiment because morality, in central part, is about putting ourselves in others' shoes and treating their interests as we do our own. When we do this at the full scale of human history, the future—where almost everyone lives and where almost all potential for joy and misery lies—comes to the fore.

This book is about *longtermism*: the idea that positively influencing the longterm future is a key moral priority of our time.[11] Longtermism is about taking seriously just how big the future could be and how high the stakes are in shaping it. If humanity survives to even a fraction of its potential life span, then, strange as it may seem, we are the ancients: we live at the very beginning of history, in the most distant past. What we do now will affect untold numbers of future people. We need to act wisely.

It took me a long time to come around to longtermism. It's hard for an abstract ideal, focused on generations of people whom we will never meet, to motivate us as more salient problems do. In high school, I worked for organisations that took care of the elderly and disabled. As an undergraduate who was concerned about global poverty, I volunteered at a children's polio rehabilitation centre in Ethiopia. When starting graduate work, I tried to figure out how people could help one another more effectively. I committed to donating at least 10 percent of my income to charity, and I cofounded an organization, Giving What We Can, to encourage others to do the same.[12]

These activities had a tangible impact. By contrast, the thought of trying to improve the lives of unknown future people initially left me cold. When a colleague presented me with arguments for taking the long term seriously, my immediate reaction was glib dismissal. There are real problems in the world facing real people, I thought, problems like extreme poverty, lack of education, and death from easily preventable diseases. That's where we should focus. Sci-fi-seeming speculations about what might or might not impact the future seemed like a distraction.

But the arguments for longtermism exerted a persistent force on my mind. These arguments were based on simple ideas: that, impartially considered, future people should count for no less, morally, than the present generation; that there may be a huge number of future people; that life, for them, could be extraordinarily good or inordinately bad; and that we really can make a difference to the world they inhabit.

The most important sticking point for me was practical: Even if we should care about the longterm future, what can we do? But as I learned more about the potentially history-shaping events that could occur in the near future, I took more seriously the idea that we might soon be approaching a critical juncture in the human story. Technological development is creating new threats and opportunities for humanity, putting the lives of future generations on the line.

I now believe the world's long-run fate depends in part on the choices we make in our lifetimes. The future could be wonderful: we could create a flourishing and long-lasting society, where everyone's lives are better than the very best lives today. Or the future could be terrible, falling to authoritarians who use surveillance and AI to lock in their ideology for all time, or even to AI systems that seek to gain power rather than promote a thriving society. Or there could be no future at all: we could kill ourselves off with biological weapons or wage an all-out nuclear war that causes civilisation to collapse and never recover.

There are things we can do to steer the future onto a better course. We can increase the chance of a wonderful future by improving the values that guide society and by carefully navigating the development of AI. We can ensure we get a future at all by preventing the creation or use of new weapons of mass destruction and by maintaining peace between the world's great powers. These are challenging issues, but what we do about them makes a real difference.

So I shifted my priorities. Still unsure about the foundations and implications of longtermism, I switched my research focus and cofounded two organisations to investigate these issues further: the Global Priorities Institute at Oxford University, and the Forethought Foundation. Drawing on what I have learned, I have tried to write the case for longtermism that would have convinced me a decade ago.

To illustrate the claims in this book, I rely on three primary metaphors throughout. The first is of humanity as an imprudent teenager. Most of a teenager's life is still ahead of them, and their decisions can have lifelong impacts. In choosing how much to study, what career to pursue, or which risks are too risky, they should think not just about short-term thrills but also about the whole course of the life ahead of them.

The second is of history as molten glass. At present, society is still malleable and can be blown into many shapes. But at some point, the glass might cool, set, and become much harder to change. The resulting shape could be beautiful or deformed, or the glass could shatter altogether, depending on what happens while the glass is still hot.

The third metaphor is of the path towards longterm impact as a risky expedition into uncharted terrain. In trying to make the future better, we don't know exactly what threats we will face or even exactly where we are trying to go; but, nonetheless, we can prepare ourselves. We can scout out the landscape ahead of us, ensure the expedition is well resourced and well coordinated, and, despite uncertainty, guard against those threats we are aware of.

This book's scope is broad. Not only am I arguing for longtermism; I'm also trying to work out its implications. I've therefore relied heavily on an extensive team of consultants and research assistants. Whenever I've stepped outside of moral philosophy, my area of expertise, domain experts have advised me from start to end. This book is therefore not really "mine": it has been a team effort. In total, this book represents over a decade's worth of full-time work, almost two years of which was spent fact-checking.

For those who want to dig deeper into some of my claims, I have compiled extensive supplementary materials, including special reports I commissioned as background research, and made them available at whatweowethefuture.com. Despite the work done so far, I believe we have only scratched the surface of longtermism and its implications; there is much still to learn.

If I'm right, then we face a huge responsibility. Relative to everyone who could come after us, we are a tiny minority. Yet we hold the entire future in our hands. Everyday ethics rarely grapples with such a scale. We need to build a moral worldview that takes seriously what's at stake.

By choosing wisely, we can be pivotal in putting humanity on the right course. And if we do, our great-great-grandchildren will look back and thank us, knowing that we did everything we could to give them a world that is just and beautiful.

CHAPTER 1

The Case for Longtermism

The Silent Billions

Future people count. There could be a lot of them. We can make their lives go better.

This is the case for longtermism in a nutshell. The premises are simple, and I don't think they're particularly controversial. Yet taking them seriously amounts to a moral revolution—one with far-reaching implications for how activists, researchers, policy makers, and indeed all of us should think and act.

Future people count, but we rarely count them. They cannot vote or lobby or run for public office, so politicians have scant incentive to think about them. They can't bargain or trade with us, so they have little representation in the market. And they can't make their views heard directly: they can't tweet, or write articles in newspapers, or march in the streets. They are utterly disenfranchised.

Previous social movements, such as those for civil rights and women's suffrage, have often sought to give greater recognition and influence to disempowered members of society. I see longtermism as an extension of these ideals. Though we cannot give genuine political power to future people, we can at least give consideration to them. By abandoning the tyranny of the present over the future, we can act as trustees—helping to create a flourishing world for generations to come. This is of the utmost importance. Let me explain why.

Future People Count

The idea that future people count is common sense. Future people, after all, are people. They will exist. They will have hopes and joys and pains and regrets, just like the rest of us. They just don't exist *yet.*

To see how intuitive this is, suppose that, while hiking, I drop a glass bottle on the trail and it shatters. And suppose that if I don't clean it up, later a child will cut herself badly on the shards.[1] In deciding whether to clean it up, does it matter *when* the child will cut herself? Should I care whether it's a week, or a decade, or a century from now? No. Harm is harm, whenever it occurs.

Or suppose that a plague is going to infect a town and kill thousands. You can stop it. Before acting, do you need to know when the outbreak will occur? Does that matter, just on its own? No. The pain and death at stake are worthy of concern regardless.

The same holds for good things. Think of something you love in your own life; maybe it's music or sports. And now imagine someone else who loves something in their life just as much. Does the value of their joy disappear if they live in the future? Suppose you can give them tickets to see their favourite band or the football team they support. To decide whether to give them, do you need to know the delivery date?

Imagine what future people would think, looking back at us debating such questions. They would see some of us arguing that future people don't matter. But they look down at their hands; they look around at their lives. What is different? What is less real? Which side of the debate will seem more clear-headed and obvious? Which more myopic and parochial?

Distance in time is like distance in space. People matter even if they live thousands of miles away. Likewise, they matter even if they live thousands of years hence. In both cases, it's easy to mistake distance for unreality, to treat the limits of what we can see as the limits of the world. But just as the world does not stop at our doorstep or our country's borders, neither does it stop with our generation, or the next.

These ideas are common sense. A popular proverb says, "A society grows great when old men plant trees under whose shade they will never sit."[2] When we dispose of radioactive waste, we don't say, "Who cares if this poisons people centuries from now?" Similarly, few of us who care about climate change or pollution do so solely for the sake of people alive today. We build museums and parks and bridges that we hope will last for generations; we invest in schools and longterm scientific projects; we preserve paintings, traditions, languages; we protect beautiful places. In many cases, we don't draw clear lines between our concerns for the present and the future—both are in play.

Concern for future generations is common sense across diverse intellectual traditions. The *Gayanashagowa*, the centuries-old oral constitution of the Iroquois Confederacy, has a particularly clear statement. It exhorts the Lords of the Confederacy to "have always in view not only the present but also the coming generations."[3] Oren Lyons, a faithkeeper for the Onondaga and Seneca nations of the Iroquois Confederacy, phrases this in terms of a "seventh-generation" principle, saying, "We . . . make every decision that we make relate to the welfare and well-being of the seventh generation to come. . . . We consider: will this be to the benefit of the seventh generation?"[4]

However, even if you grant that future people count, there's still a question of how much weight to give their interests. Are there reasons to care more about people alive today?

Two reasons stand out to me. The first is partiality. We often have stronger special relationships with people in the present, like family, friends, and fellow citizens, than with people in the future. It's common sense that you can and should give extra weight to your near and dear.

The second reason is reciprocity. Unless you live as a recluse in the wilderness, the actions of an enormous number of people—teachers, shopkeepers, engineers, and indeed all taxpayers—directly benefit you and have done so throughout your life. We typically think that if someone has benefited you, that gives you a reason to repay them. But future people don't benefit you the way others in your generation do.[5]

Special relationships and reciprocity are important. But they do not change the upshot of my argument. I'm not claiming that the interests of present and future people should always and everywhere be given equal weight. I'm just claiming that future people matter significantly. Just as caring more about our children doesn't mean ignoring the interests of strangers, caring more about our contemporaries doesn't mean ignoring the interests of our descendants.

To illustrate, suppose that one day we discover Atlantis, a vast civilisation at the bottom of the sea. We realise that many of our activities affect Atlantis. When we dump waste into the oceans, we poison its citizens; when a ship sinks, they recycle it for scrap metal and other parts. We would have no special relationships with the Atlanteans, nor would we owe them repayment for benefits they had bestowed on us. But we should still give serious consideration to how our actions affect them.

The future is like Atlantis. It, too, is a vast, undiscovered country;[6] and whether that country thrives or falters depends, in significant part, on what we do today.

The Future Is Big

It's common sense that future people count. So, too, is the idea that, morally, the numbers matter. If you can save one person or ten from dying in a fire, then, all else being equal, you should save ten; if you can cure a hundred people or a thousand of a disease, you should cure a thousand. This matters, because the number of future people could be huge.

To see this, consider the long-run history of humanity. There have been members of the genus *Homo* on Earth for over 2.5 million years.[7] Our species, *Homo sapiens*, evolved around three hundred thousand years ago. Agriculture started just twelve thousand years ago, the first cities formed only six thousand years ago, the industrial era began around 250 years ago, and all the changes that have happened since then—transitioning from horse-drawn

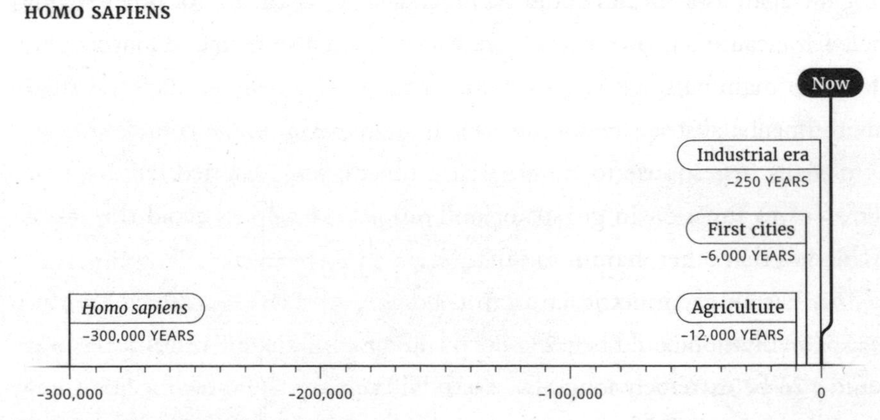

Figure 1.1. The history of Homo sapiens.

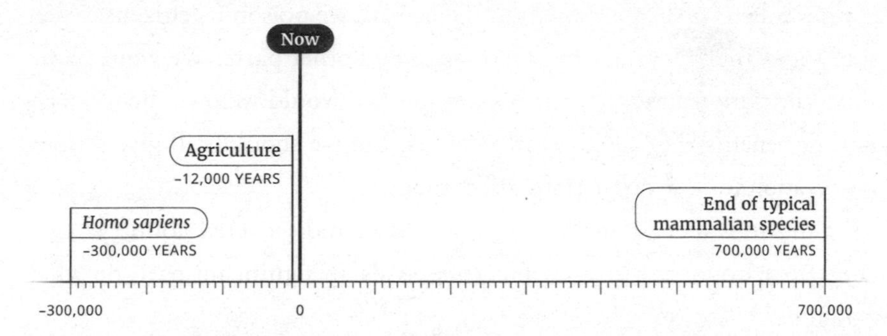

Figure 1.2. The potential future of civilisation, if humans survive as long as the average mammalian species.

carts to space travel, leeches to heart transplants, mechanical calculators to supercomputers—occurred over the course of just three human lifetimes.[8]

How long will our species last? Of course, we don't know. But we can make informative estimates that take our uncertainty into account, including our uncertainty about whether we'll cause our own demise.

To illustrate the potential scale of the future, suppose that we only last as long as the typical mammalian species—that is, around one million years.[9] Also assume that our population continues at its current size. In that case, there would be eighty trillion people yet to come; future people would outnumber us ten thousand to one.

Of course, we must consider the whole range of ways the future could go. Our life span as a species could be much shorter than that of other mammals if we cause our own extinction. But it could also be much longer. Unlike other mammals, we have sophisticated tools that help us adapt to varied environments; abstract reasoning, which allows us to make complex, long-term plans in response to novel circumstances; and a shared culture that allows us to function in groups of millions. These help us avoid threats of extinction that other mammals can't.[10]

This has an asymmetric impact on humanity's life expectancy. The future of civilisation could be very short, ending within a few centuries. But it could also be extremely long. The earth will remain habitable for hundreds of millions of years. If we survive that long, with the same population per

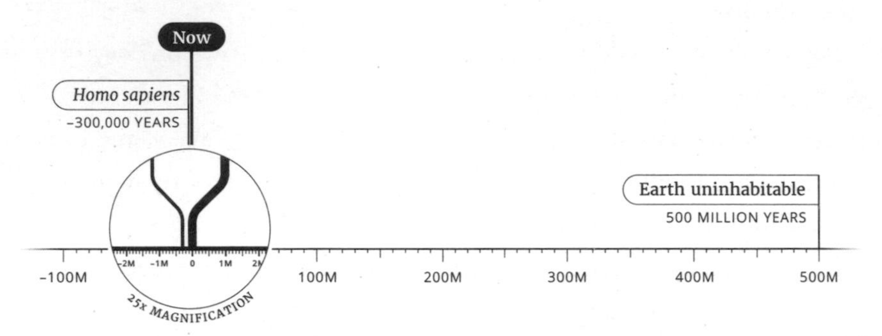

Figure 1.3. The potential future of civilisation if it survives until the earth becomes uninhabitable for humans due to the sun's increasing brightness. There is considerable uncertainty as to the length of this window, with estimates ranging from 500 million to 1.3 billion years.

century as now, there will be a million future people for every person alive today. And if humanity ultimately takes to the stars, the timescales become literally astronomical. The sun will keep burning for five billion years; the last conventional star formations will occur in over a trillion years; and, due to a small but steady stream of collisions between brown dwarfs, a few stars will still shine a million trillion years from now.[11]

The real possibility that civilisation will last such a long time gives humanity an enormous life expectancy. A 10 percent chance of surviving five hundred million years until the earth is no longer habitable gives us a life expectancy of over fifty million years; a 1 percent chance of surviving until the last conventional star formations give us a life expectancy of over ten billion years.[12]

Ultimately, we shouldn't care just about humanity's life expectancy but also about how many people there will be. So we must ask: How many people in the future will be alive at any one time?

Future populations might be much smaller or much larger than they are today. But if the future population is smaller, it can be smaller by eight billion at most—the size of today's population. In contrast, if the future population is bigger, it could be much bigger. The current global population is already over a thousand times larger than it was in the hunter-gatherer era. If global population density increased to that of the Netherlands—an agricultural net exporter—there would be seventy billion people alive at any one time.[13] This might seem fantastical, but a global population of eight billion would have seemed fantastical to a prehistoric hunter-gatherer or an early agriculturalist.

Population size could get dramatically larger again if we one day take to the stars. Our sun produces billions of times as much sunlight as lands on Earth, there are tens of billions of other stars across our galaxy, and billions of galaxies are accessible to us.[14] There might therefore be vastly more people in the distant future than there are today.

Just how many? Precise estimates are neither possible nor necessary. On any reasonable accounting, the number is immense.

To see this, look at the following diagram. Each figure represents ten billion people. So far, roughly one hundred billion people have ever lived. These past people are represented as ten figures. The present generation consists of almost eight billion people, which I'll round up to ten billion and represent with a single figure:

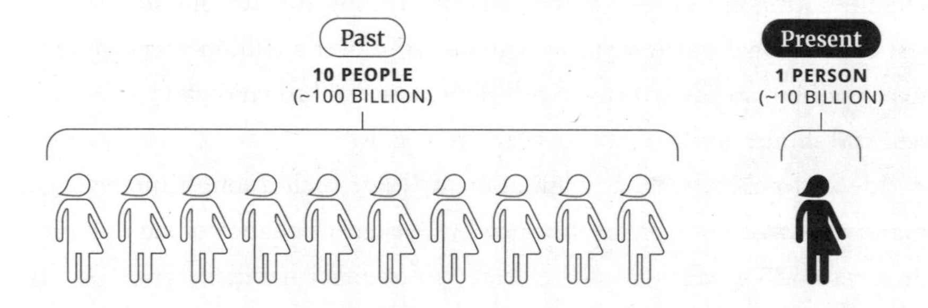

Next, we'll represent the future. Let's just consider the scenario where we stay at current population levels and live on Earth for five hundred million years. These are all the future people:

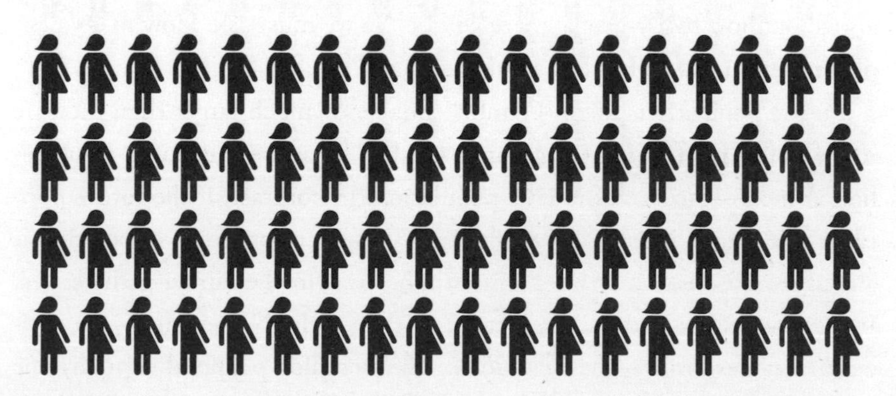

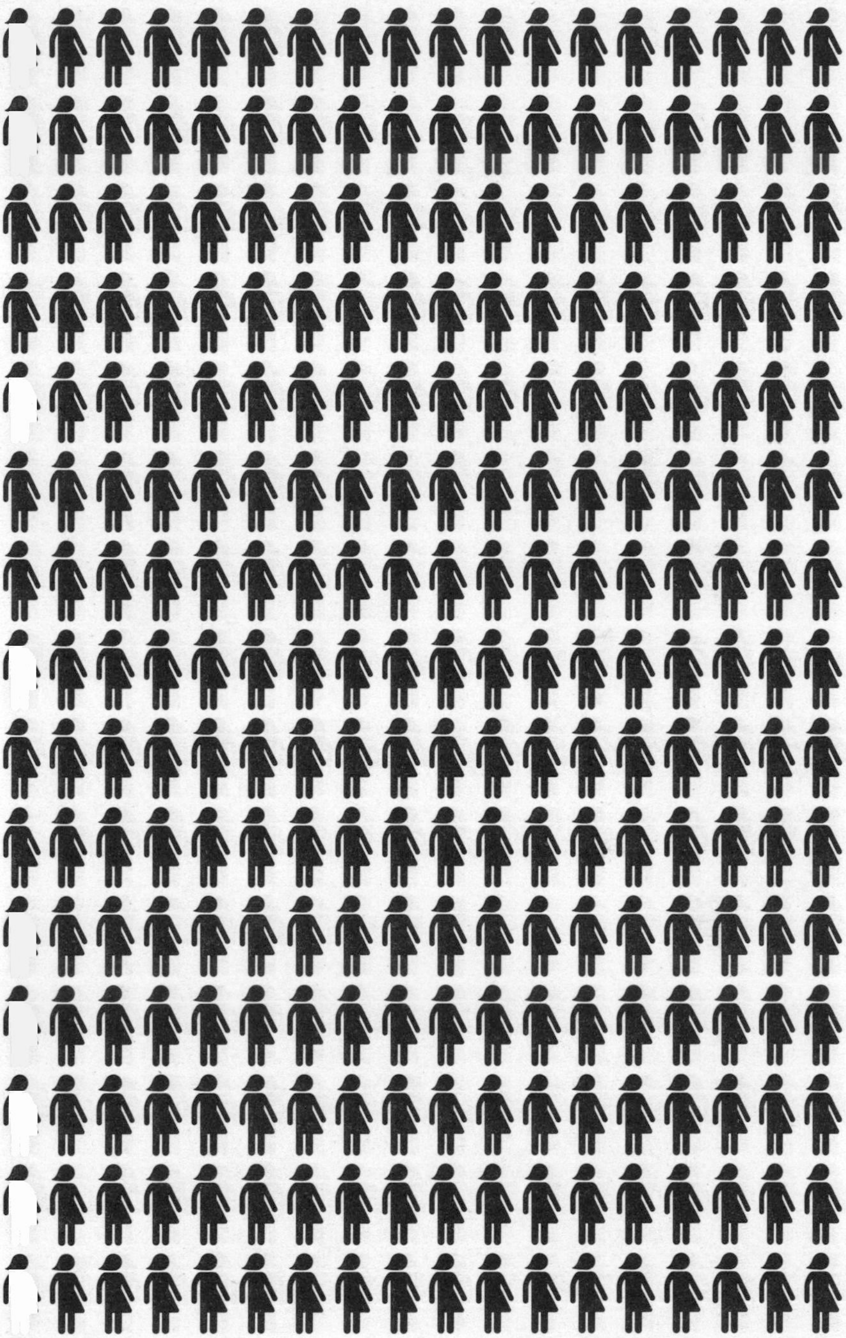

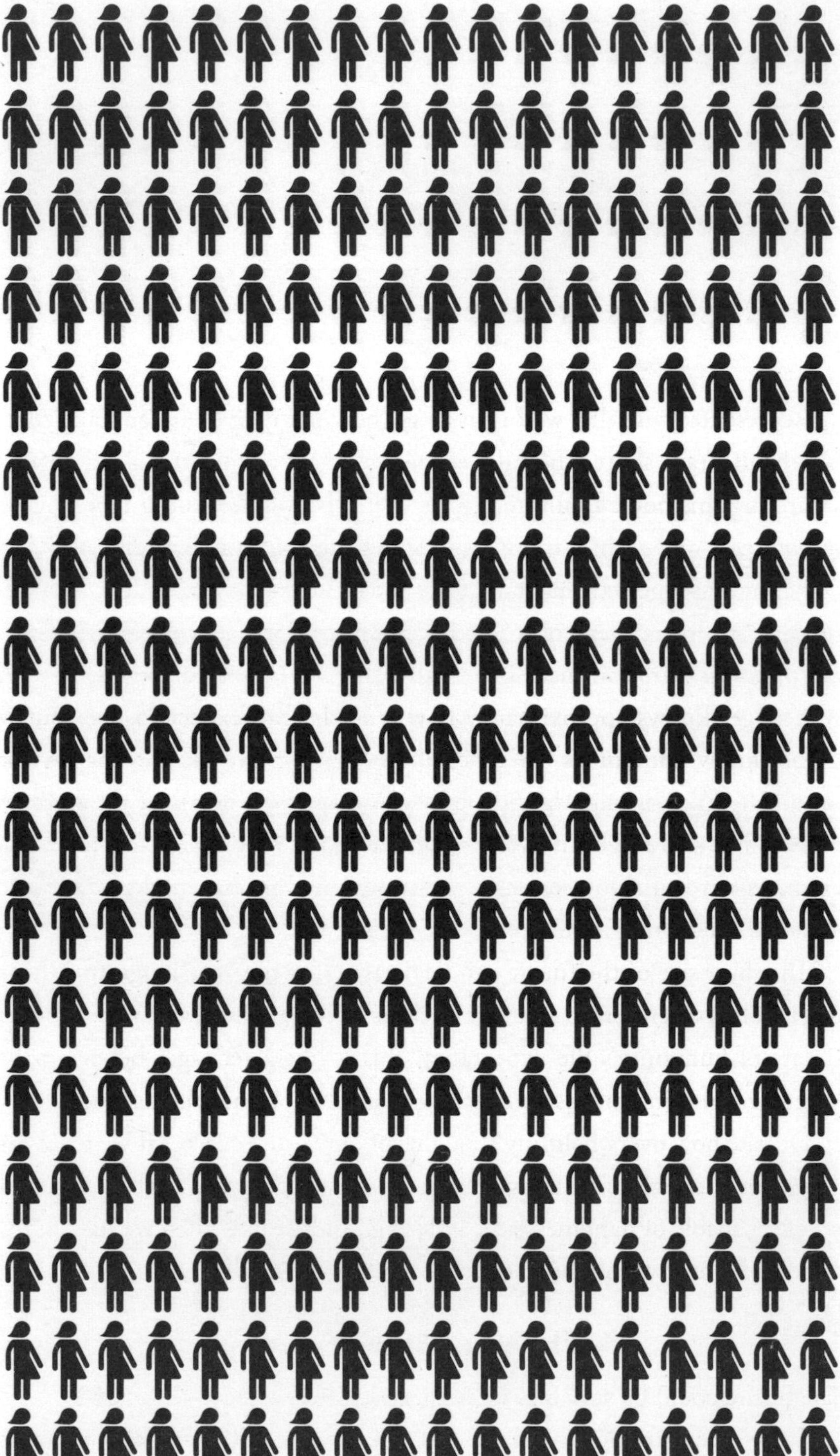

Represented visually, we begin to see how many lives are at stake. But I cut the diagram short. The full version would fill twenty thousand pages—saturating this book a hundred times over. Each figure would represent ten billion lives, and each of those lives could be flourishing or wretched.

Earlier, I suggested that humanity today is like an imprudent teenager: most of our life is ahead of us, and decisions that impact the rest of that life are of colossal importance. But, really, this analogy understates my case. A teenager knows approximately how long she can expect to live. But we do not know humanity's life expectancy. We are more like a teenager who, for all she knows, might accidentally cause her own death in the next few months but also might live for a thousand years. If you were in such a situation, would you think seriously about the long life that might be ahead of you, or would you ignore it?

The sheer size of the future can be dizzying. Typically, "longterm" thinking involves attention to years or decades at most. But even with a low estimate of humanity's life expectancy, this is like a teenager believing that longterm thinking means considering tomorrow but not the day after.

Despite how overwhelming thoughts of our future can be, if we truly care about the interests of future generations—if we recognize that they are real people, capable of happiness and suffering just like us—then we have a duty to consider how we might impact the world they inhabit.

The Value of the Future

The future could be very big. It could also be very good—or very bad.

To get a sense of how good, we can look at some of the progress humanity has made over the last few centuries. Two hundred years ago, average life expectancy was less than thirty; today, it is seventy-three.[15] Back then, over 80 percent of the world lived in extreme poverty; now, less than 10 percent does.[16] Back then, only about 10 percent of adults could read; today, more than 85 percent can.[17]

Collectively we have the power both to encourage these positive trends and to change course on the negative trends, like the dramatic increases in carbon dioxide emissions and in the number of animals suffering in factory farms. We can build a world where everyone lives like the happiest people in the most well-off countries today, a world where no one lives in poverty, no one lacks adequate medical care, and, insofar as is possible, everyone is free to live as they want.

But we could do even better still—far better. The best that we have seen so far is a poor guide to what is possible. To get some inkling of this, consider the life of a rich man in Britain in 1700—a man with access to the best food, health care, and luxuries available at the time. For all his advantages, such a man could easily die of smallpox, syphilis, or typhus. If he needed surgery or had a toothache, the treatment would be agonising and carry a significant risk of infection. If he lived in London, the air he breathed would be seventeen times as polluted as it is today.[18] Travelling even within Britain could take weeks, and most of the globe was entirely inaccessible to him. If he had imagined a future merely where most people were as rich as him, he would have failed to anticipate many of the things that improve our lives, like electricity, anaesthesia, antibiotics, and modern travel.

It's not just technology that has improved people's lives; moral change has done so, too. In 1700, women were unable to attend university, and the feminist movement did not exist.[19] If that well-off Brit was gay, he could not love openly; sodomy was punishable by death.[20] In the late 1700s, three in four people globally were the victims of some form of forced labour; now less than 1 percent are.[21] In 1700, no one lived in a democracy. Now over half the world does.[22]

Much of the progress we've made since 1700 would have been very difficult for people back then to anticipate. And that's with only a three-century gap. Humanity could last for millions of centuries on Earth alone. On such a scale, if we anchor our sense of humanity's potential to a fixed-up version of our present world, we risk dramatically underestimating just how good life in the future could be.

Consider the very best moments in your life—moments of joy, beauty, and energy, like falling in love, or achieving a lifelong goal, or having some creative insight. These moments provide proof of what is possible: we know that life can be at least as good as it is then. But they also show us a direction in which our lives can move, leading somewhere we have yet to go. If my best days can be hundreds of times better than my typically pleasant but humdrum life, then perhaps the best days of those in the future can be hundreds of times better again.

I'm not claiming that a wonderful future is *likely*. Etymologically, "utopia" means "no-place," and indeed the path from here to some ideal future state is very fragile. But a wonderful future is not just a fantasy, either. A better word would be "eutopia," meaning "good place"—something to strive for. It's a future that, with enough patience and wisdom, our descendants could actually build—if we pave the way for them.

And though the future could be wonderful, it could also be terrible. To see this, look at some of the negative trends of the past and imagine a future where *they* are the dominant forces guiding the world. Consider that slavery had all but disappeared from France and England by the end of the twelfth century, but in the colonial era those same countries became slave traders on a massive scale.[23] Or consider that the mid-twentieth century saw totalitarian regimes emerging even out of democracies. Or that we used scientific advances to build nuclear weapons and factory farms.

Just as eutopia is a real possibility, so is dystopia. The future could be one where a single totalitarian regime controls the world, or where today's quality of life is but a distant memory of a former Golden Age, or where a third world war has led to the complete destruction of civilisation.

Not Just Climate Change

Even if you accept that the future is big and important, you might be skeptical that we can positively affect it. And I agree that working out the long-run effects of our actions is very hard. There are many considerations at play, and our understanding of them is just beginning. My aim with this book is to stimulate further work in this area, not to be definitive in any conclusions about what we should do. But the future is so important that we've got to at least try to figure out how to steer it in a positive direction. And, already, there are some things we can say.

Looking to the past, though there are not many examples of people deliberately aiming at long-run impacts, they do exist, and some had surprising levels of success. Poets provide one source. In Shakespeare's Sonnet 18 ("Shall I compare thee to a summer's day?") the author notes that through his art he can preserve the young man he admires for all eternity:[24]

But thy eternal summer shall not fade,
. . . .
When in eternal lines to time thou grow'st.
So long as men can breathe or eyes can see,
So long lives this, and this gives life to thee.[25]

Sonnet 18 was written in the 1590s but echoes a tradition that goes back much further.[26] In 23 BC the Roman poet Horace began the final poem in his *Odes* with these lines:[27]

> I have finished a monument more lasting than bronze, more lofty than the regal structure of the pyramids, one which neither corroding rain nor the ungovernable North Wind can ever destroy, nor the countless series of the years, nor the flight of time.
>
> I shall not wholly die, and a large part of me will elude the Goddess of Death.[28]

These claims seem bombastic, to say the least. But, plausibly, these poets' attempts at immortality succeeded. They have survived many hundreds of years and are in fact flourishing as the years pass: more people read Shakespeare today than did in his own time, and the same is probably true of Horace. And as long as some member of each future generation is willing to pay the tiny cost involved in preserving or replicating some representation of these poems, they will persist forever.

Other writers have also successfully aimed at very longterm impact. Thucydides wrote his *History of the Peloponnesian War* in the fifth century BC.[29] Many consider him the first Western historian to try to depict events faithfully and analyse their causes.[30] He believed he was describing general truths, and he deliberately wrote his history so that it could be influential far into the future:

> It will be enough for me, however, if these words of mine are judged useful by those who want to understand clearly the events which happened in the past and which (human nature being what it is) will, at some time or other and in much the same ways, be repeated in the future. My work is not a piece of writing designed to meet the taste of an immediate public, but was done to last for ever.[31]

Thucydides's work is still enormously influential to this day. It is required reading at the West Point and Annapolis military academies and the US Naval War College.[32] The widely read 2017 book *Destined for War*, by political scientist Graham Allison, had the subtitle *Can America and China Escape Thucydides's Trap?* Allison analyses US-China relations in the same terms that Thucydides used for Sparta and Athens. As far as I know, Thucydides is the first person in recorded history to have deliberately aimed at longterm impact and succeeded.

More recent examples come from the United States' Founding Fathers. The US Constitution is almost 250 years old and has mostly remained the same throughout its life. Its founding was of enormous longterm importance, and many of the Founding Fathers were well aware of this. John Adams, the second president of the United States, commented, "The institutions now made in America will not wholly wear out for thousands of years. It is of the last importance, then, that they should begin right. If they set out wrong, they will never be able to return, unless it be by accident, to the right path."[33]

Similarly, Benjamin Franklin had such a reputation for believing in the health and longevity of the United States that in 1784 a French mathematician wrote a friendly satire of him, suggesting that if Franklin was sincere in his beliefs, he should invest his money to pay out on social projects centuries later, getting the benefits of compound interest along the way.[34] Franklin thought it was a great idea, and in 1790 he invested £1000 (about $135,000 in today's money) each for the cities of Boston and Philadelphia: three-quarters of the funds would be paid out after one hundred years, and the remainder after two hundred years. By 1990, when the final funds were distributed, the donation had grown to almost $5 million for Boston and $2.3 million for Philadelphia.[35]

The Founding Fathers themselves were influenced by ideas developed almost two thousand years before them. Their views on the separation of powers were foreshadowed by Locke and Montesquieu, who drew on Polybius's analysis of Roman governance from the second century BC.[36] We also know that several Founding Fathers were familiar with Polybius's work themselves.[37]

Those of us in the present don't need to be as influential as Thucydides or Franklin to predictably impact the longterm future. In fact, we do it all the time. We drive. We fly. We thereby emit greenhouse gases with very long-lasting effects. Natural processes will return carbon dioxide concentrations to preindustrial levels only after hundreds of thousands of years.[38] These are timescales usually associated with radioactive nuclear waste.[39] However, with nuclear power we carefully store and plan to bury the waste products; with fossil fuels we belch them into the air.[40]

In some cases, the geophysical impacts of this warming get even more extreme over time rather than "washing out."[41] The Intergovernmental Panel on Climate Change (IPCC) projects that in the medium-low-emissions scenario, which is now widely seen to be the most likely, sea level would rise by around 0.75 metres by the end of the century.[42] But it would keep rising well past the year 2100. After ten thousand years, sea level would be ten to twenty metres higher than it is today.[43] Hanoi, Shanghai, Kolkata, Tokyo, and New York would all be mostly below sea level.[44]

Climate change shows how actions today can have longterm consequences. But it also highlights that longterm-oriented actions needn't involve ignoring the interests of those alive today. We can positively steer the future while improving the present, too.

Moving to clean energy has enormous benefits in terms of present-day human health. Burning fossil fuels pollutes the air with small particles that

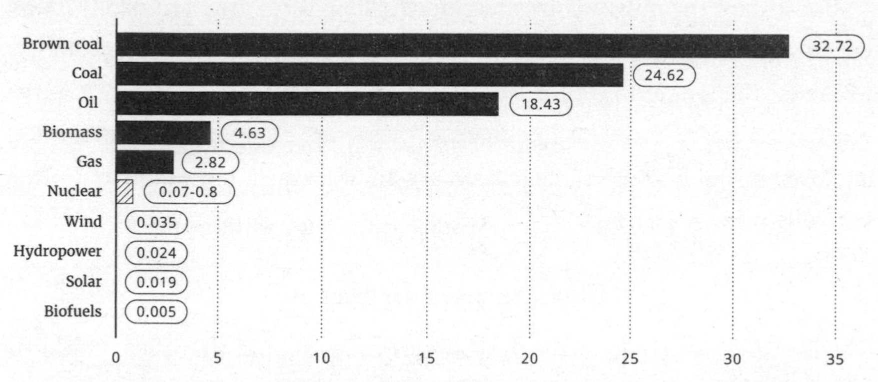

Figure 1.4. Deaths per terawatt-hour of produced electricity for various power sources; includes both deaths from accidents and from air pollution, but not from contributions to climate change. The nuclear power figure includes the accidents at Chernobyl and Fukushima; the displayed range is due to differing estimates of the longer-term effects of low-radiation exposure—for more detail, see whatweowethefuture.com/notes. Estimates for other power sources are based on data from Europe.

cause lung cancer, heart disease, and respiratory infections.[45] As a result, every year about 3.6 million people die prematurely.[46] Even in the European Union, which in global terms is comparatively unpolluted, air pollution from fossil fuels causes the average citizen to lose a whole year of life.[47]

Decarbonisation—that is, replacing fossil fuels with cleaner sources of energy—therefore has large and immediate health benefits in addition to the longterm climate benefits. Once one accounts for air pollution, rapidly decarbonising the world economy is justified by the health benefits alone.[48]

Decarbonisation is therefore a win-win, improving life in both the long and the short term. In fact, promoting innovation in clean energy—such as solar, wind, next-generation nuclear, and alternative fuels—is a win on other fronts, too. By making energy cheaper, clean energy innovation improves living standards in poorer countries. By helping keep fossil fuels in the ground, it guards against the risk of unrecovered collapse that I'll discuss in Chapter 6. By furthering technological progress, it reduces the risk of longterm stagnation that I'll discuss in Chapter 7. A win-win-win-win-win.

Decarbonisation is a proof of concept for longtermism. Clean energy innovation is so robustly good, and there is so much still to do in that area that I see it as a baseline longtermist activity against which other potential actions can be compared. It sets a high bar.

But it's not the only way of affecting the long term. The rest of this book tries to give a systematic treatment of the ways in which we can positively influence the longterm future, suggesting that moral change, wisely governing the ascent of artificial intelligence, preventing engineered pandemics, and averting technological stagnation are all at least as important, and often radically more neglected.

Our Moment in History

The idea that we could affect the longterm future, and that there could be so much at stake, might just seem too wild to be true. This is how things initially seemed to me.[49]

But I think that the wildness of longtermism comes not from the moral premises that underlie it but from the fact that we live at such an unusual time.[50]

We live in an era that involves an extraordinary amount of change. To see this, consider the rate of global economic growth, which in recent decades averaged around 3 percent per year.[51] This is historically unprecedented. For the first 290,000 years of humanity's existence, global growth was close to 0 percent per year; in the agricultural era that increased to around 0.1 percent, and it accelerated from there after the Industrial Revolution. It's only in the last hundred years that the world economy has grown at a rate above 2 percent per year. Putting this another way: from 10,000 BC onwards, it took

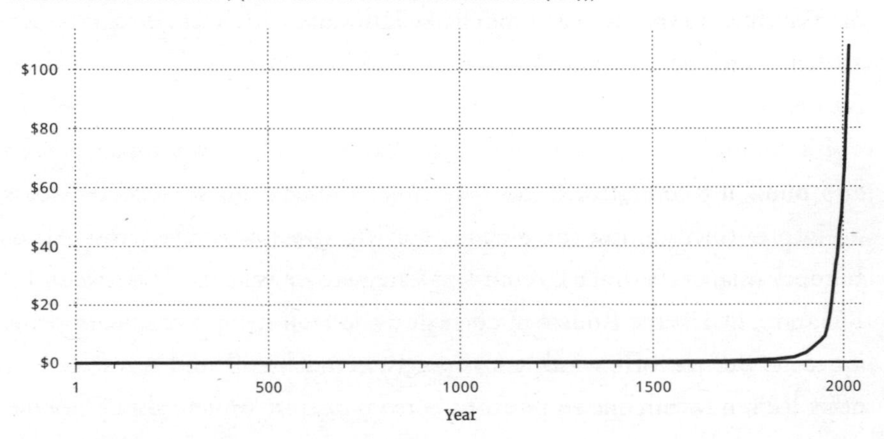

Figure 1.5. World economic output since AD 1.

many hundreds of years for the world economy to double in size. The most recent doubling took just nineteen years.[52] And it's not just that rates of economic growth are historically unusual; the same is true for rates of energy use, carbon dioxide emissions, land use change, scientific advancement, and arguably moral change, too.[53]

So we know that the present era is extremely unusual compared to the past. But it's also unusual compared to the future. This rapid rate of change cannot continue forever, even if we entirely decouple growth from carbon emissions and even if in the future we spread to the stars. To see this, suppose that future growth slows a little to just 2 percent per year.[54] At such a rate, in ten thousand years the world economy would be 10^{86} times larger than it is today—that is, we would produce one hundred trillion trillion trillion trillion trillion trillion trillion times as much output as we do now. But there are less than 10^{67} atoms within ten thousand light years of Earth.[55] So if current growth rates continued for just ten millennia more, there would have to be ten million trillion times as much output as our current world produces for *every atom* that we could, in principle, access. Though of course we can't be certain, this just doesn't seem possible.[56]

Humanity might last for millions or even billions of years to come. But the rate of change of the modern world can only continue for thousands of years. What this means is that we are living through an extraordinary chapter in humanity's story. Compared to both the past and the future, every decade we live through sees an extremely unusual number of economic and technological changes. And some of these changes—like the inventions of fossil fuel power, nuclear weapons, engineered pathogens, and advanced artificial intelligence—have the potential to impact the whole course of the future.

It's not only the rapid rate of change that makes this time unusual. We're also unusually connected.[57] For over fifty thousand years, we were broken up into distinct groups; there was simply no way for people across Africa, Europe, Asia, or Australia to communicate with one another.[58] Between 100 BC and AD 150 the Roman Empire and the Han dynasty each comprised up to 30 percent of the world's population, yet they barely knew of each other.[59] Even within one empire, one person had very limited ability to communicate with someone far away.

In the future, if we spread to the stars, we will again be separated. The galaxy is like an archipelago, vast expanses of emptiness dotted with tiny pinpricks of warmth. If the Milky Way were the size of Earth, our solar system would be ten centimetres across and hundreds of metres would separate us from our neighbours. Between one end of the galaxy and the other, the fastest possible communication would take a hundred thousand years; even between us and our closest neighbour, there-and-back communication would take almost nine years.[60]

In fact, if humanity spreads far enough and survives long enough, it will eventually become impossible for one part of civilisation to communicate with another. The universe is composed of millions of groups of galaxies.[61] Our own is called, simply, the Local Group. The galaxies within each group are close enough to each other that gravity binds them together forever.[62] But, because the universe is expanding, the groups of galaxies will eventually be torn apart from each other. Over 150 billion years in the future, not even light will be able to travel from one group to another.[63]

The fact that our time is so unusual gives us an outsized opportunity to make a difference. Few people who ever live will have as much power to positively influence the future as we do. Such rapid technological, social, and environmental change means that we have more opportunity to affect when and how the most important of these changes occur, including by managing technologies that could lock in bad values or imperil our survival. Civilisation's current unification means that small groups have the power to influence the whole of it. New ideas are not confined to a single continent, and they can spread around the world in minutes rather than centuries.

The fact that these changes are so recent means, moreover, that we are out of equilibrium: society has not yet settled down into a stable state, and we are able to influence *which* stable state we end up in. Imagine a giant ball rolling rapidly over a rugged landscape. Over time it will lose momentum and slow, settling at the bottom of some valley or chasm. Civilisation is like this ball: while still in motion, a small push can affect in which direction we roll and where we come to rest.

CHAPTER 2

You Can Shape the Course of History

Prehistory's Impact on Today

Human beings have been making choices with longterm consequences for tens of thousands of years. Consider: Why is Africa home to so many more species of megafauna—large animals like elephants and giraffes—than the rest of the world?[1] You might think, as I did before learning about this topic, that the answer has to do with Africa's particular environment. But that's not right. Fifty thousand years ago, a great variety of megafauna roamed the planet.

Consider the glyptodonts, a group of armadillo-like herbivores that lived in South America for tens of millions of years.[2] The largest glyptodonts were as big and heavy as cars.[3] Their bodies were encased in a giant shell, they had a bone helmet, and some of them had club-shaped tails adorned with spikes.[4] They looked like giant capybaras dressed up as armoured trucks. They went extinct around 12,000 years ago.[5]

Or consider megatherium, a giant ground sloth and one of the largest land mammals to have ever lived, rivalling the Asian elephant in size.[6] It went extinct 12,500 years ago.[7] Or *Notiomastodon*, a genus of elephant-like animals with giant tusks that evolved two million years ago and went extinct 10,000 years ago.[8] Or the dire wolf, the largest known canine to have lived, which, having lost its giant herbivorous prey, went extinct 13,000 years ago.[9] All these species lived in South America, along with dozens of other megafauna species that are no longer with us.

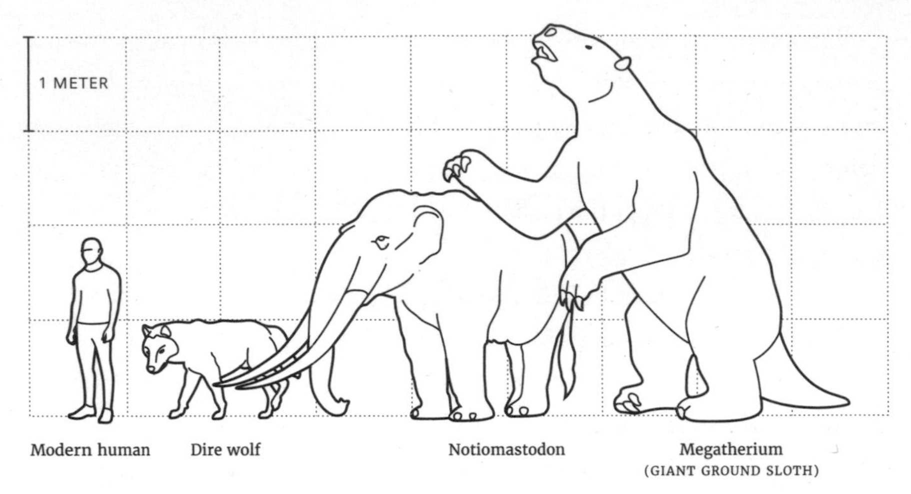

Figure 2.1. Some specimens of now-extinct megafauna drawn at scale in comparison with a modern human.

There is a heated debate over what caused the extinctions of megafauna. Some scientists believe that natural climate change was the main driver, some believe that humans were the culprit, and some believe it was a mix of humans and climate change.[10] In my view, the evidence is clear that humans often played a decisive role: most of these megafauna survived over a dozen similarly sized climatic changes in the past;[11] smaller animals did not go extinct at nearly the same rate as megafauna;[12] and the timing of their extinction usually coincides with humans' arrival into their habitats.[13] Though perhaps helped by climate change, it was hunting and the disruption of natural environments caused by human activity that killed them off. Unlike megafauna on other continents, African megafauna evolved alongside humans and so were better prepared for *Homo sapiens* as a predator.

The extinction of these megafauna was probably an irrevocable change to the world, made by humans with extremely primitive technology. It meant we lost, for all time, many beautiful and unique species. And *Homo sapiens* are not only implicated in the extinction of giant sloths and canines: we are also the prime suspect in the end of our human cousins, the Denisovans and the Neanderthals, who likely died out as a result of both competition and interbreeding.[14] There is now only one *Homo* species on the planet, but there could have been many.

Early humans made other choices with longterm consequences, too. Early agriculturalists, for example, burned down vast swathes of forest to create plains for farming and paddies for rice irrigation.[15] This preindustrial deforestation had a lasting impact. Because carbon dioxide remains in the atmosphere for so long, the planet is, as a result of the actions of our ancestors, slightly warmer today.[16]

Just as actions taken by our ancestors thousands of years ago shaped the present day, so too will decisions we make today shape the future thousands of years hence. But to justify taking a longterm view of our decisions, what matters is not only whether we can impact the future but whether we can adequately foresee what those impacts will be. We don't need to predict every detail, nor could we if we tried. But if we want to make the future better, we need to identify actions that have positive effects on balance over very long timescales.

Our distant ancestors could not predict their longterm impact on the world. Hunter-gatherers did not know they were driving species to extinction. Early agriculturalists could not guess that deforestation would warm the planet, nor what the consequences of this warming would be.

But we in the modern era can do better. Clearly, there's still much we don't know, but in the last few centuries especially, we've learned a lot. If early agriculturalists had had our understanding of climate physics, they could have foreseen some of the geophysical impacts of burning forests; if hunter-gatherers had had our knowledge of ecology and evolutionary biology, they would have understood what it is for a species to go extinct and the potentially irrevocable loss that was at stake. With careful investigation and appropriate humility, we can now start to assess the effects of our actions over very long timescales.

In this chapter, I'll present a framework for assessing the longterm value of an event. The chapters that follow apply this framework to events that I think we, today, can foreseeably influence for the better.

A Framework for Thinking About the Future

Consider some state of affairs that people could bring about, like the nonexistence of the glyptodonts. We can assess the longterm value of this new state of affairs in terms of three factors: its significance, its persistence, and its contingency.[17]

Significance is the average value added by bringing about a certain state of affairs. How much worse is the world, at any one time, because the glyptodonts are extinct? In assessing this, we would want to attend to all relevant aspects of the glyptodonts' extinction: the intrinsic loss of a species on the planet, the loss to humans who could have used their shells or eaten their meat, and the impact on the ecosystems the glyptodonts inhabited.

The *persistence* of a state of affairs is how long that state of affairs lasts, once it has been brought about. The nonexistence of the glyptodonts may be exceptionally persistent, starting 12,000 years ago and lasting until the end of the universe.[18] It would only fail to be exceptionally persistent if, at a future time, we were to bring them back.

Technology may make this possible. There are current efforts to "de-extinct" certain species, like the woolly mammoth, by extracting DNA from their remains and editing that DNA into the cells of similar modern animals, like elephants.[19] However, even if successful, these efforts would not truly bring back the original creatures: instead, they would produce a hybrid—an animal that looks a lot like the extinct animal but is not genetically the same. Should future generations try to bring back the glyptodonts, they would probably face similar challenges.

The final aspect of the framework is *contingency*. This is the most subtle part of the framework. In English the word "contingency" has a few different meanings; in the sense I'm using it, an alternative term would be "noninevitability." Contingency represents the extent to which a state of affairs depends on a small number of specific actions. If something is very contingent, then that change would not have otherwise occurred for a very long time, or ever. The existence of the novel *Jane Eyre* is very contingent: if Charlotte Brontë had not written it, that precise novel would never have been written by someone else. Agriculture is less contingent because it emerged in multiple locations independently.

If something is very noncontingent, then the change would have happened soon anyway, even without the individual's action. Knowledge of calculus was not very contingent because Leibniz independently discovered it just a few years after Newton did. Considering contingency is crucial because if you make a change to the world but it's a change that would have simply happened soon afterward anyway, then you have not made a longterm *difference* to the world.

Though it's hard to be confident, my guess is that the extinction of the glyptodonts was not very contingent. Even if the hunters who killed off the last of them had not done so, then probably some other group of hunters, at some later time, would have. In order to prevent the glyptodonts' extinction, those hunters would have had to promote a norm that the glyptodonts should be protected and this norm would have had to be passed down the generations, and adhered to, until the present day. This would not be impossible to pull off, but it does seem difficult.

Multiplying significance, persistence, and contingency together gives us the longterm value of bringing about some state of affairs. Because of this, we can make intuitive comparisons between different longterm effects on these dimensions. For example, between two alternatives, if one is ten times as persistent as the other, that will outweigh the other being eight times as significant. Because the potential scale of the longterm future is so great—millions, billions, or even trillions of years—our attention should be, first, on what states of affairs might be the most persistent. Then, afterwards, we can think about significance and contingency.

Table 2.1. The Significance, Persistence, Contingency Framework

Significance	What's the average value added by bringing about a certain state of affairs?
Persistence	How long will this state of affairs last once it has been brought about?
Contingency	If not for the action under consideration, how briefly would the world have been in this state of affairs (if ever)?

Note: For more details, see Appendix 3.

To see how this framework can be used to guide our decisions today, let's return to the metaphor of humanity as an imprudent teenager. Looking back at our own individual teenage years, what choices mattered most? Plausibly, it's those whose effects were the most persistent, affecting the whole course of our lives; most significant, making the biggest difference to our wellbeing at any one time; and most contingent, causing an effect that would not have happened anyway at some later date.

Some choices I made as a teenager did not have persistent effects: my plans for the weekend made a difference to that weekend but usually didn't shape the course of my life. The effects of other choices were not that contingent. Like many teenagers, I cared about firsts—first drink, first time having sex. But ultimately, such firsts would have happened at some point regardless, and looking back, the precise timing did not matter much. Finally, some effects, though persistent and contingent, just weren't that significant. I chose not to get braces to close the gap between my two front teeth because at the time I believed that a gap brings good luck. I still have the gap today, but as far as I can tell, it has not significantly affected my life.

Other decisions I made mattered a lot. I was reckless as a teenager and sometimes went "buildering," also known as urban climbing. Once, coming down from the roof of a hotel in Glasgow, I put my foot on a skylight and fell through. I caught myself at waist height, but the broken glass punctured my side. Luckily, it missed all internal organs. A little deeper, though, and my guts would have popped out violently, and I could easily have died. I still have the scar: three inches long and almost half an inch thick, curved like an earthworm. Dying that evening would have prevented all the rest of my life. My choice to go buildering was therefore an enormously important (and enormously foolish) decision—one of the highest-stakes decisions I'll ever make.

More mundanely, I could easily have exposed myself to a different set of intellectual influences, which would have set me on a very different path in life. All my close friends studied medicine—the standard path for smart, socially minded teenagers in Scotland—and I considered it for myself. If I had not studied philosophy at school, and if I hadn't had such an engaged and passionate teacher, Jeremy Hall, I would probably not have studied it at university or pursued it as a career. I expect that a career in medicine would have been fulfilling, but it probably would not have exposed me to the moral arguments that led me to the path I've taken—a difference which, from my current perspective, would have been a major loss.

Looking back, it's clear that, for many of my teenage choices, what mattered most was not the fun I had at the time—whether buildering was a thrill (it was) or whether studying medicine at Edinburgh involved better parties. Rather, what mattered most was the impact of these choices on the rest of my life, whether I was risking death or altering the values that would guide my future self.

The risk of death I bore as a teenager and the intellectual influences that shaped my life mirror the two main ways in which we can impact the longterm future. First, we can affect humanity's duration: ensuring that we survive the next few centuries affects how many future generations there are. That is, we can help *ensure civilisation's survival.* Just as my teenage decisions to gamble with my life were among the most consequential I've ever made, so too are our decisions about how to handle risks of extinction or unrecovered civilisational collapse among the most consequential decisions that we as a society make today.

Second, we can affect civilisation's average value, changing how well or badly life goes for future generations, potentially for as long as civilisation lasts. That is, we can *change trajectory*, trying to improve the quality of future people's lives over the life span of civilisation.[20] Just as the intellectual influences I was exposed to as a teenager shaped the whole rest of my life, so, too, I will argue, the values that humanity adopts in the next few centuries might shape the entire trajectory of the future.[21]

These two ideas structure the book. Part II of this book looks at trajectory changes, focusing in particular on changing society's values. Within this,

TWO WAYS TO IMPROVE THE FUTURE

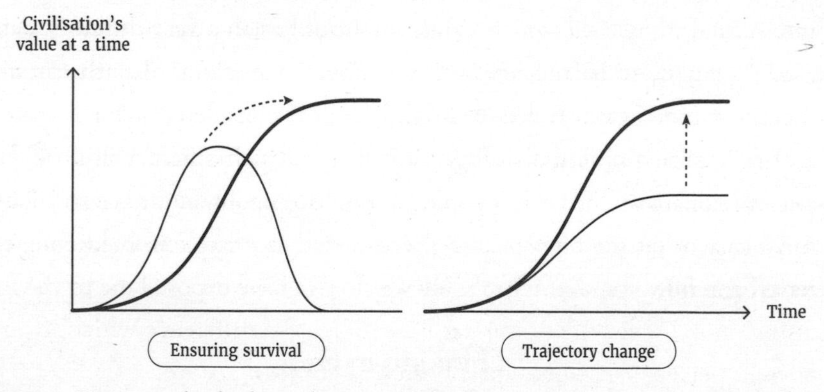

Figure 2.2. We can make the future better in two ways: by averting permanent catastrophes, thereby ensuring civilisation's survival; or by changing civilisation's trajectory to make it better while it lasts.

Chapter 3 argues for the significance and contingency of value changes, focusing on the abolition of slavery as a case study. Chapter 4 argues for the persistence of values, suggesting that new technology, in particular advanced artificial intelligence, could enable those in power to lock in their values indefinitely. Whether the future is governed by values that are authoritarian or egalitarian, benevolent or sadistic, exploratory or rigid, might well be determined by what happens this century.

Part III looks at three ways of ensuring survival, dedicating a chapter to each. The first way is to prevent direct risks of human extinction; I focus on engineered pandemics. The second is to prevent the unrecovered collapse of civilisation; I focus on risks from nuclear war and extreme climate change. The third is technological stagnation, which could increase the risks of both extinction and collapse. Along the way, I discuss the persistence and contingency of the end of civilisation.

The question of the *significance* of the end of civilisation raises philosophical issues. Broadly, ensuring survival increases the quantity of future life; trajectory changes increase its quality. But you might not care much about sheer quantity. If there's no longer anyone around to care, why should it matter if civilisation has ended? And maybe, on balance, the future is more bad than good. If these worries were correct, then the longtermist priority should be to increase the average value of future civilisation rather than its duration. Improving our trajectory would be more important than ensuring survival.

Part IV tackles these issues. I argue both that we should think of the nonexistence of future generations as a moral loss, if the people in them would have sufficiently good lives, and that we should expect the future to be more good than bad, on balance. Ensuring survival is therefore just as great a priority as improving our trajectory.

Part V turns to action. Longtermism is not just abstract philosophical speculation. It's an idea that people are putting into practice today. Chapter 10 looks at what some people are doing today to try to make the long term better, and how you can help.[22]

Thinking in Bets

When thinking about the changes that we could make to the world, we will not know how long they will last or how significant or contingent they will be. So we need a way of making decisions in the face of uncertainty. The most widely accepted account of how to do so is expected value theory.

Over the course of writing this book, I was repeatedly and viscerally reminded of the idea of expected value theory by my housemate at the time, Liv Boeree. Liv is one of the most successful female poker players of all time—a European Poker Tour and World Series champion. Her understanding and internalisation of the idea of expected value—or "EV," as she calls it—is critical to her success.

There are three aspects to expected value. First, probabilities. Rather than thinking that a three-of-a-kind poker hand is "very unlikely," Liv knows that the chance of getting one, before any cards are dealt, is about 5 percent; if the first two cards she's dealt are a pair, this probability rises to about 12 percent.[23] Though both probabilities are small, the difference between them can easily be enough to affect your decisions at the poker table.

What's striking about Liv is that she applies this same probabilistic thinking to other areas of her life, too. She and her partner, Igor (another poker player), will happily discuss the probability that they'll still be together after ten years. (It's currently at 80 percent.)

It can feel unnatural to apply probabilities to areas of life where chances aren't easily quantified. But it means we can have more nuanced and accurate views about the world. It's a way of thinking more precisely. "People often think something definitely will or definitely won't happen—as zero percent chance or a hundred percent chance," Liv told me. "But of course almost everything falls in between. Or else they use vague language like 'a fair chance.' But a 'fair chance' means very different things to different people."

She's right. One study found that people interpret the phrase "might happen" to refer to anything between 10 percent and 60 percent probability, and "a serious possibility" as all the way from 30 percent to 90 percent.[24] This vagueness can have momentous implications. In 1961, when President John F. Kennedy asked the military for advice on whether to invade Cuba at the Bay of Pigs, he was told that the plan had a "fair chance" of success. Quite reasonably, Kennedy took that to be a positive assessment. But the author of the words "fair chance" later said that he meant that there was only about a 30 percent chance of success.[25] The operation failed dramatically.

The second aspect of expected value is assigning values to outcomes. For professional poker players, this is comparatively easy: they can just look at their financial returns. But financial returns are not in general the right measure of value. If you need £1000 to pay for a life-saving operation, then the difference in value for you between getting nothing and getting £1000 is much greater than the difference in value between getting £1000 and getting £2000. The value that we assign to outcomes should be based on whatever it is we *ultimately* care about, such as people's wellbeing.

Precisely assigning value to different outcomes can be difficult, but we often only need very rough comparisons in order to make a decision. Suppose that there are two different drugs that could cure a patient's ailment, with different side effects. The first will certainly cause a mild headache; the second has a one-in-ten risk of causing a fatal heart attack. It's hard to know exactly how much worse death is than a mild headache. But, apart from exceptional cases, it's certainly more than ten times worse.

This brings us to the third aspect of expected value theory, which is measuring how good or bad a decision is by its expected value. This can be intuitive: in the two-drugs example I just gave, the first drug is the better choice; death is more than ten times as bad as a mild headache, so a 10 percent risk of death is sufficient to outweigh a guarantee of a headache. We can calculate the expected value of a decision as follows. First, we list each possible outcome of the decision. Next, we assign a probability and a value to each outcome, which we then multiply together. Finally, we add up all the probability-times-value products.

Liv and Igor make bets against each other all the time, and they decide whether to take them on the basis of expected value. To take one real-life example, suppose that Liv and Igor are at a pub, and Liv bets Igor that he can't flip and catch six coasters at once with one hand. If he succeeds, she'll give him £3; if he fails, he has to give her £1. Suppose Igor thinks there's a fifty-fifty chance that he'll succeed. If so, then it's worth it for him to take the bet: the upside is a 50 percent chance of £3, worth £1.50; the downside is a 50 percent chance of losing £1, worth negative £0.50. Igor makes an expected £1 by taking the bet—£1.50 minus £0.50. If his beliefs about his own chances of success are accurate, then if he were to take this bet over and over again, on average he'd make £1 each time.

Table 2.2. Igor's Decision

	Catches the coasters (50% probability)	Fails to catch the coasters (50% probability)	Expected payoff
Take bet	£3	–£1	£1
Refuse bet	£0	£0	£0

Expected value theory is not just useful when gambling. It's crucial whenever we have to take a bet—that is, to make a decision in the face of uncertainty—which is almost all the time. My teenage decisions make this vivid. Before going buildering, I dismissed the possibility of falling and dying as unlikely and therefore not worth worrying about. But that was hugely foolish—not because it was *likely* that I would fall and die, but because it wasn't *sufficiently unlikely*, and dying is so bad that even a small chance is well worth avoiding.

In the face of an uncertain future, humanity often acts like my reckless teenage self. For example, climate change sceptics often point to our uncertainty as a reason for inaction.[26] There's so much we don't know, they claim—we don't know exactly how well climate models predict the amount of warming for a given quantity of emissions, for instance, or just how damaging a certain amount of warming would be for the economy. So we should not waste resources on the problem. But this is a terrible argument. We can grant that there's great uncertainty about what climate change means. But uncertainty cuts both ways. The damage caused by climate change might be less than is typically forecasted, but it might also be considerably *worse*—if, for example, the climate is more sensitive to temperature changes than such forecasts presuppose, or adaptation is harder, or we will emit more carbon dioxide than experts currently predict.

Crucially, the uncertainty around climate change is not symmetric: greater uncertainty should prompt more concern about worst-case outcomes, and this shift is not offset by a higher chance of best-case outcomes, because the worst-case outcomes are worse than the best-case outcomes are good.[27] For example, according to the Intergovernmental Panel on Climate Change, on the medium-low-emissions scenario, the best guess is that we will end up with around 2.5 degrees Celsius of warming by the end of the century.[28] But this is uncertain. There is a one-in-ten chance that we get 2 degrees or less. But that should not reassure us, because there is also a one-in-ten chance that we get more than 3.5 degrees.[29] Less than 2 degrees would be something of a relief compared to the best-guess estimate, but more than 3.5 degrees would be much worse. The uncertainty gives us *more* reason to worry, not less. It's as if my teenage self, before jumping off a building, had reassured onlookers by saying, "It's OK, I've no idea how far I'll fall!"

Much the same will be true for the issues that I cover in this book. I'm not saying that we should be confident that value lock-in or major catastrophe will occur this century. What I am saying is that their chance of occurring is very real—certainly more than 1 percent, and certainly greater than many everyday risks, like dying in a car crash. When combined with how much is at stake, the expected value of trying to ensure a good future is enormous.

When we're applying the significance, persistence, and contingency framework, we should therefore be thinking about expected significance, expected persistence, and expected contingency.[30] If some change to the world has an 80 percent chance of fizzling out after ten years but a 20 percent chance of lasting for a million years, then its expected persistence is over two hundred thousand years. In general, if some change to the world has at least a reasonable chance of being highly significant, persistent, and contingent, then that can be sufficient for the expected value of that change to be very great indeed.

Moments of Plasticity

Often, some event can have highly significant, persistent, and contingent effects if there is a period of plasticity, where ideas or events or institutions can take one of many forms, followed by a period of rigidity or ossification. The dynamic is like that of glassblowing: In one period, the glass is still molten and malleable; it can be blown into one of many shapes. After it cools, it becomes rigid, and further change is impossible without remelting.

Plasticity frequently comes after a crisis, like a war. For example, after the end of World War II, Korea was divided along the thirty-eighth parallel. The location of the division was extremely contingent. Colonel Dean Rusk and Charles Bonesteel, two American officers in their midthirties using a

National Geographic map, proposed the thirty-eighth parallel because it divided the country roughly in half while keeping Seoul on the American side.[31] They were working on short notice because the United States had to reach an agreement with the Soviet Union before the entire peninsula fell into Soviet hands. No experts on Korea were consulted, and the proposed border cut across several preexisting Korean provinces and geographic features. In fact, the United States was surprised that the Soviets accepted the division; not only did it give Seoul to the United States, but Soviet troops were already in Korea while the closest American forces were still in Okinawa, several hundred miles away.[32] Yet after the division was implemented, it became hard to reverse, and it has since resulted in enormous differences to the fates of those who ended up in each of those two countries. South Koreans live in a strong democracy and are almost thirty times richer on average than they were in 1953. North Koreans live under a totalitarian dictatorship and may be even poorer than they were before the Korean War.[33]

A period of plasticity also commonly occurs when some idea or institution is still new. For example, the US Constitution was written over just four months—a moment of great plasticity—and amended eleven times in its first six years of operation.[34] After that, though, it became more rigid. Between 1804 and 1913, only three amendments were passed, all immediately following the Civil War: they abolished slavery, granted citizenship to African Americans and formerly enslaved people, and prohibited race from influencing the right to vote.[35] Today, the Constitution is again very rigid: it's only been amended once in the last fifty years, and that amendment—to prevent increases in congressional salaries from taking effect until the next term of office—was first proposed in 1789.[36]

This dynamic can hold for the laws and norms relevant to new technologies, too. Following World War II, the international community debated a variety of ways nuclear weapons could be governed.[37] One proposal, put forward by the United States, was the Baruch Plan, according to which the United States would disband its nuclear weapons programme and transfer its bombs to the UN to be destroyed. The UN would then oversee the mining of fissionable materials around the world and inspect other countries to ensure that no one was building nuclear bombs. The USSR countered with the Gromyko Plan, which also proposed universal disarmament. Both of these plans failed, and it's not clear that either ever had much of a chance. But it was clearly a time of much greater plasticity in nuclear governance than we see now. Today, the idea that the UN could control the mining of uranium seems entirely off the table.

The dynamic of "early plasticity, later rigidity" can hold for new ideas, too. In addition to the books that we now know as the New Testament, a number of other texts were taught by some early Christians.[38] The New Testament books became the core Christian teachings only over the course of the first and second centuries AD and were not cemented until around the end of the fourth century AD.[39]

A final example comes from the history of climate change activism. The effect that carbon dioxide would have on global warming was first quantified in 1896 by Svante Arrhenius; his 1906 estimate of equilibrium climate sensitivity was four degrees, which is only a little higher than modern estimates.[40] And it was knowable, at that time, that we would probably emit dramatically more carbon dioxide in the future: one simply needed to continue extrapolating the trend of exponential economic growth and to recognize the obvious fact that such growth would bring a corresponding increase in energy demand.

In 1958, Frank Capra, director of *It's a Wonderful Life*, made an educational weather documentary, *Unchained Goddess*, which included a warning about climate change: "Even now, man may be unwittingly changing the world's climate through the waste products of his civilisation. Due to our release through factories and automobiles every year of more than six billion tonnes of carbon dioxide, which helps air absorb heat from the sun, our atmosphere seems to be getting warmer. . . . [It's] been calculated that a few degrees rise in the earth's temperature would melt the polar ice caps."[41] Two years earlier, referencing work by Gilbert Plass, the *New York Times* had published an article arguing that carbon dioxide emissions were warming the planet. As with Svante Arrhenius's, Plass's estimate of equilibrium climate sensitivity—3.6 degrees—was strikingly close to the Intergovernmental Panel on Climate Change's current best estimate.[42]

If we had taken action on climate change earlier, we would have been acting on more speculative evidence than we have now. But the issue would also have been much less politically divisive, and change might have been much easier. Bill McKibben, one of the world's leading environmentalists, suggested this, saying in 2019: "Thirty years ago, there were relatively small things we could have done that would have changed the trajectory of this battle—a small price on carbon back then would have yielded a different trajectory, would have put us in a different place. We might not have solved climate change yet because it's a huge problem, but we'd be on the way."[43]

The lesson Bill McKibben takes from the history of climate change activism is that we should pay close attention to new challenges as they arise. He highlights advanced artificial intelligence in particular: "We haven't taken [advanced artificial intelligence] seriously because it doesn't, at the moment, impinge on our day-to-day life. But one of the things that climate change taught me is that things happen fast, like, really fast. And, before you know it, they're out of control. So the time for thinking about them is when there is still some chance of getting a handle on them."[44] He's right. With climate change, we may have missed one moment of plasticity, and we should hope there are more to come. But perhaps we can also learn a more general lesson and respond more rapidly to new challenges—like artificial intelligence, synthetic biology, tensions between the United States and China, the rise of new ideologies, and the potential slowdown in technological progress—as soon as they arise. These are some of the issues I'll cover in the next two parts of this book.

Indeed, over the next two chapters, I'll suggest that the dynamic of "early plasticity, later rigidity" could be true for history as a whole. We are currently in a period where the values that guide civilisation are still malleable, but I'll argue in Chapter 4 that, within the next few centuries, those values could ossify, constraining the course of all future civilisation. If so, then changes we make to today's moral values could have indefinitely long-lasting impacts. Let's turn to this idea, focusing first on the *contingency* of moral change.

PART II
TRAJECTORY CHANGES

CHAPTER 3

Moral Change

Abolition

Despite its abhorrence, slavery was almost ubiquitous historically.[1] In one form or another, slavery was practised across Europe, Africa, the Americas, and Asia. It existed in almost all early agricultural civilisations, including ancient Mesopotamia, Egypt, China, and India.[2] People were enslaved for a variety of reasons: as a result of conquest or kidnapping, because of inability to repay debts, as punishment for crimes, or because their family sold them.[3] In the Roman Empire, probably at least 10 percent of the population was enslaved.[4] The Arab world, stretching from modern-day Morocco to modern-day Oman, also had a long-standing and extensive slave trade that lasted until the twentieth century. People were bought or raided from Africa, Central Asia, and Christian Europe and typically forced to work as soldiers or personal servants, or enslaved for sex.[5] Estimates vary, but in total about twelve million people were enslaved in Africa alone in the trans-Saharan and Indian Ocean slave trades.[6]

Slave trading reached its apogee in the transatlantic trade, fuelled by Europeans' desire to exploit abundant land and natural resources in the Americas. Over twelve million enslaved people were taken from Africa, including 470,000 to British North America, 1.6 million to the Spanish colonies, 4.2 million to the Caribbean, and 5.5 million to Brazil.[7] Though Europeans sometimes enslaved people by raiding, most often they bought them from African leaders who had enslaved them from other communities.[8]

The conditions in transit across the Atlantic were abominable. Enslaved people were packed into transport ships in cramped, poorly ventilated

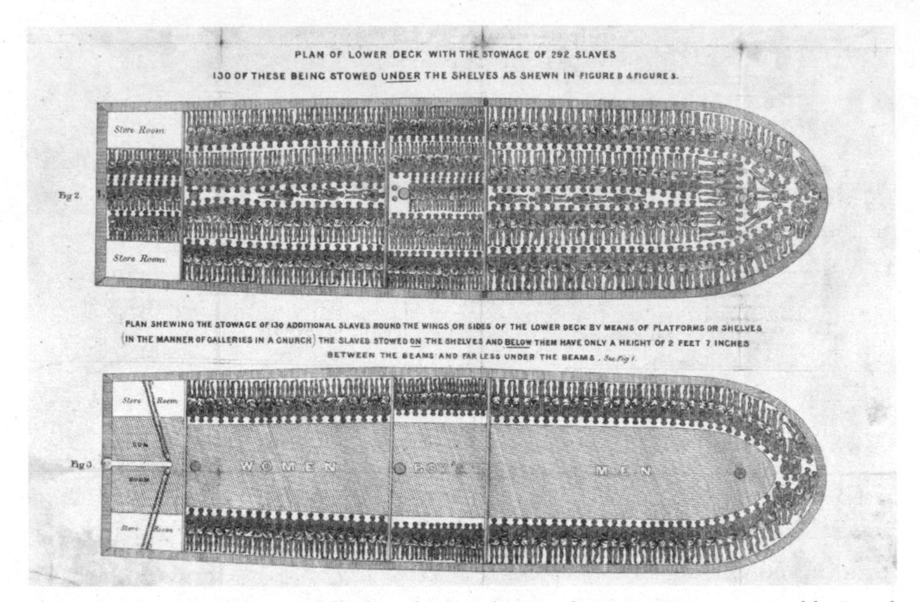

Figure 3.1. A 1780s diagram of the slave ship Brookes, *used as campaigning material by British abolitionists.*

quarters. Disease was rampant.[9] The enslaved were forbidden from using the ship's toilets and were forced to lie in their own feces for weeks. Around 1.5 million people died on these voyages.[10]

The suffering of those who survived the journey across the Atlantic is impossible to accurately convey. The enslaved were typically forced to work on plantations—most often those growing sugar cane, tobacco, cotton, or coffee—and sometimes to mine silver or gold.[11] Work days were regularly ten hours long, and pregnant women and children were sometimes also forced to work.[12] By 1700, enslaved people made up the overwhelming majority of the population of the Caribbean, and their life expectancy at birth was sometimes as low as twenty years.[13] Although most British colonies had codes that regulated treatment of the enslaved, in practice slave owners acted as judge, jury, and executioner. Whipping was widespread as a means of disincentivizing "inefficient labor" and keeping enslaved people in a state of fear.[14]

It's hard to imagine how people could believe that owning other people was permissible. We might naturally think that slave owners really knew, deep down, that what they were doing was wrong and that they didn't care. But we should be careful not to presume that the values of other people are more similar to our own than they really are. Slavery was seen as entirely permissible, part of the natural order.[15] Historically, even thinkers who dedicated their lives to moral reflection, often highly progressive in other areas, accepted slavery. These included the classical philosophers Plato and Aristotle, and Enlightenment thinkers such as Immanuel Kant.[16]

Yet despite its historical ubiquity, its longevity, and its acceptance, and despite the luminaries who defended it, slavery was abolished. Was its abolition inevitable, a result of economic changes or the inexorable march of moral progress? Or was it a contingent matter, where if history had gone down a different path, it might never have occurred?

A full account of abolition would require a book in its own right and would cover the countless acts of resistance, subversion, and bravery by enslaved people throughout history.[17] It would also cover efforts from formerly enslaved people such as Frederick Douglass, Sojourner Truth, and Harriet Tubman in the United States and Luís Gama in Brazil, who shed light on the horrors of slavery, fostered public opposition, and pushed for legislative action.

Here, though, I look at just one part of this narrative. Because I'm interested in whether or not abolition was contingent, I'm interested in those parts of the history that seem unexpected or difficult to explain. And, as leading historian of abolition Professor Christopher Leslie Brown puts it, "The causes of slave resistance do not seem particularly mysterious."[18] What is surprising, he notes, is that slavery was attacked by those who benefited from it. Moreover, enslaved people have very often throughout history powerfully resisted their oppression. So why was there a successful abolitionist campaign in Britain in the early 1800s and not in any of history's previous slave societies?

I think that the activism of a fairly small group of Quakers in the eighteenth and early nineteenth centuries provides part of the answer. Their efforts were hugely important in one of the most surprising moral about-faces in history. There were many important figures in this story, but among the early Quaker activists, the most striking was Benjamin Lay.[19]

Lay was born in Copford, England, in 1682. He became a sailor based in London, then a shopkeeper in Barbados, before moving in 1732 to Philadelphia, which at the time was the largest city in British North America and home to the largest Quaker community. Lay was a dwarf, standing at a little over four feet tall, and hunchbacked. He referred to himself as "Little Benjamin," likening himself to "little David" who killed Goliath.[20]

Lay's moral radicalism took many forms. He opposed the death penalty and consumerism.[21] Like many of the later abolitionists, and very unusually for the time, he became a vegetarian and even refused to wear leather or wool. Later in his life, he lived in a cave just outside Philadelphia and, boycotting all goods produced by enslaved people, made all his own clothes, wore undyed fabrics, and refused to drink tea or eat sugar.[22]

His opposition to slavery stemmed from his time as a sailor, when he learned of the pervasiveness of rape on the transatlantic slave ships, and from the two years he spent in Barbados. Early in his time there, he whipped several enslaved people who, racked by hunger, had stolen food from his shops. He was subsequently stricken with guilt and made friends with a number of enslaved people.[23] One of these friends, a barrel maker, had a master who would whip the people he owned every Monday morning "to keep them in awe."[24] One Sunday evening, in order to avoid the next day's brutality, this friend committed suicide. Experiences like these haunted Lay for the rest of his life.

Over the course of the twenty-seven years that he lived in Pennsylvania, Lay harangued the Philadelphia Quakers about the horrors of slavery at every opportunity, and he did so in dramatic style. He once stood outside a Quaker meeting in the snow in bare feet with no coat. When passersby expressed concern, he explained that enslaved people were made to work outside for the whole winter dressed as he was. During Quaker meetings, as soon as any slave owner tried to speak, it was said that Lay would rise to his feet and shout, "There's another negro-master!"[25] When kicked out of one meeting for making trouble, he lay down in the mud outside the entrance of the meetinghouse so that every member of the congregation had to step over his body as they left.[26] When he discovered that a local family kept a young girl as a slave, he invited their six-year-old son to his cave without telling his parents so that they would briefly know the grief of losing a child.[27]

In his most famous stunt, at the 1738 Yearly Meeting of the Quakers, he came dressed in military uniform under a large cloak, carrying a hollow book filled with fake blood. During the meeting, he allegedly rose to his feet, threw off his cloak, and exclaimed, "Oh all you Negro masters who are contentedly holding your fellow creatures in a state of slavery, . . . you might as well throw off the plain coat as I do. It would be as justifiable in the sight of the Almighty, who beholds and respects all nations and colours of men with an equal regard, if you should thrust a sword through their hearts as I do through this book!"[28] As he spoke, he splattered the gathering with the fake blood. John Woolman, who later became one of the most influential Quaker abolitionists, was likely in the audience that day.[29]

Lay became well known across Pennsylvania.[30] But he was not revered in his time for his activism. In fact, he was effectively disowned four times, by Quaker societies in London, Colchester, Philadelphia, and Abington.[31] But he seems, ultimately, to have been influential within Quaker circles: in the late 1790s, Benjamin Rush wrote that a print of Lay was seen in "many houses in Philadelphia."[32] Lay was also friends with Anthony Benezet, who helped to make abolition mainstream in Britain.[33] And Lay's activism coincided with the time when moral sentiment among Quakers changed dramatically. In the period of 1681 to 1705, an estimated 70 percent of the leaders of the Quaker's Yearly Meeting owned people; for the period 1754 to 1780 that figure was only 10 percent.[34] In the 1758 Philadelphia Yearly Meeting, it was decided that Quakers who traded people would be disciplined and then disowned (though it would be another eighteen years before *owning* people was also banned).[35] When Lay was told, he reportedly shouted, "Thanksgiving and praise be rendered unto the Lord God. . . . I can now die in peace."[36] He passed away one year later.

One can find buds of abolitionist thought throughout history. Enslaved people themselves frequently and often violently objected to the inhumane treatment they suffered. Moralists occasionally condemned slavery's cruelties, sometimes worrying about its effect on the enslavers as well as the enslaved.[37] They recommended treating enslaved people better or releasing them as a matter of charity or for religious reasons.[38] Many were uneasy about how the institution could coexist with certain tenets of their faith or, for various eighteenth-century Enlightenment thinkers, with the principles of universalism or natural rights.[39] In practical terms, some rulers occasionally tried to increase the freedom of their subjects in order to curtail the power of their nobles or prevent uprisings.[40] But the Quakers seem to be the first group in history to organize a campaign for abolition, push for public support, and seek to stamp out slavery entirely.[41]

The activism of Lay and others inspired a generation of abolitionists who provided a crucial bridge between North American Quaker thought and mass appeal in Britain. Anthony Benezet was particularly influential. He founded a school for young Black people in 1770 to demonstrate that they were as intellectually capable as White people.[42] Many of the students, such as Absalom Jones, Richard Allen, and James Forten, went on to become leading campaigners for abolition themselves.[43] Benezet's work also inspired Thomas Clarkson, a cofounder of the Society for Effecting the Abolition of the Slave Trade, to take up the cause. Clarkson in turn convinced the parliamentarian William Wilberforce to become the political leader of the British abolitionist movement.[44]

Working together with formerly enslaved people such as Olaudah Equiano and Ottobah Cugoano, who formed the Sons of Africa—Britain's first Black political organization[45]—the abolitionists' campaign in Britain was enormously successful. Britain's parliament was persuaded to abolish the slave trade in 1807 and to make *owning* people illegal across most of the British Empire in 1833.[46] After 1807 the British government resolved to stamp out slave trading worldwide. They used diplomacy and bribery to persuade other nations to ban the transatlantic slave trade and used the Royal Navy's West Africa Squadron to police the seas.[47] This made it harder for slave ships to travel between West Africa, the United States, and the American and Caribbean colonies of France, Spain, Portugal, and Holland. The campaign ultimately captured more than two thousand slave ships and freed over two hundred thousand enslaved people, although those freed were often exploited in other ways and sent to work across the British Empire.[48]

The abolition of slavery was an example of a *values change*, by which I mean a change in the moral attitudes of a society, or in how those attitudes are implemented and enforced. In my view, the abolition of slavery was one of the most important values changes in all of history. Over the course of this chapter and the next, I'll argue that changing society's values is particularly important from a longtermist perspective. This chapter will look at the significance and contingency of values changes; the next chapter will discuss their persistence.

The Significance of Values

The significance of a state of affairs is how good or bad it is at any point in time. The example of slavery makes the significance of values changes obvious. Abolition freed millions of people from lives of utter misery. But it is far from the only example of the extreme significance of moral values.

Consider moral views on the status of women. Throughout history, women have been systematically oppressed. In 1832, twenty-five years after it abolished the slave trade, the British government passed the Great Reform Act to officially prohibit women from voting. Today, women can vote in every democracy in the world and have far greater opportunities to work and participate in public life. But since attitudes regarding gender roles still vary widely across different countries, some women have more opportunities than others. For example, Cambodia, Laos, Vietnam, India, and Pakistan all have about the same income per capita. But in Cambodia, Laos, and Vietnam, about three out of every four women participate in the labour force, while in India and Pakistan fewer than one in four do.[49]

Other examples abound. In the last few decades, attitudes towards LGBTQ+ people have changed dramatically in many countries. The first US state to legalize gay marriage was Massachusetts, in 2004. Just eleven years later, a Supreme Court decision legalized it nationwide. As a result of these changing attitudes, millions of people are now more able to live full, enfranchised lives.

Corporal punishment in schools, widespread throughout much of the twentieth century, is now prohibited in more than 120 countries.[50] Evolving attitudes towards nationalism and immigration have life-changing implications for the hundreds of millions of international migrants;[51] one estimate found that, on average, for a low-skill worker, moving to the United States boosts their annual income by over $15,000 per year.[52] And it's not only people who are affected by our values. Landscapes and ecosystems can be reshaped by the extent to which we value nature. Our attitudes towards animal welfare have huge implications for the billions of animals that are raised in factory farms.[53]

Values changes are significant because they have major impacts on the lives of people and other beings. But from a longtermist perspective, they are particularly significant compared to other sorts of changes we might make because their effects are unusually predictable.

If you promote a particular means of achieving your goals, like a particular policy, you run the risk that the policy might not be very good at achieving your goal in the future, especially if the world in the future is very different from today, with a very different political, cultural, and technological environment. You might also lose out on the knowledge that we will gain in the future, which might change whether we even think that this policy is a good idea. In contrast, if you can ensure that people in the future adopt a particular *goal*, then you can trust them to pursue whatever strategies make the most sense, in whatever environment they are in and with whatever additional information they have. You can therefore be fairly confident that you have made the achievement of that goal more likely, even if you have no idea at all what the world will be like when those future people act.

The "dead hand problem" in philanthropy illustrates the importance of promoting goals rather than means. Often the founders of a charity specify a constitution that directs the future behaviour of that charity in ways that become absurd over time. One example is ScotsCare—"the charity for Scots in London"—which is dedicated to improving the lives of Scottish Londoners. This particular goal made sense at the time of the charity's founding in 1611. At that time, Scotland and England had only recently come under the rule of the same king; Scots in London were immigrants, and some were unusually deprived and unable to receive support from their local parish, the equivalent of social security at the time.[54] But this goal makes less sense four hundred years later. London is the most affluent city in the UK,[55] and as far as I can tell, Scots nowadays face no particular disadvantages there. In contrast, many areas within Scotland are far more deprived. Presumably the founders of the charity did not care about Scots in London per se; they just cared about their fellow nationals. They would have done better at achieving their aims if they had directed the charity to pursue the goal they fundamentally cared about—"Do whatever will best improve the lives of Scots"—rather than mandating a very particular way of reaching that goal.

For these reasons changing values has particularly great significance from a longterm perspective. Looking to the past, we see that such changes have had an enormous impact on the lives of billions of people. Looking to the future, if we can improve the values that guide the behaviour of generations to come, we can be pretty confident that they will take better actions, even if they're living in a world very different from our own, the nature of which we cannot predict.

The Contingency of Values

However, if some change we make to society's values would simply have happened anyway, then the long-run impact of that change is not so great. So we also need to consider the *expected contingency* of values changes. We need to ask, If we don't bring about some change to society's values, how long (in expectation) would it take for that change to happen anyway? Today we say the abolitionist movement had a crucial role in ending slavery. But if, for some reason, abolition was inevitable, then over the long run the changes the abolitionists fought for would have happened anyway, at some later date.

Contingency can vary depending on the timescale we're considering. It's more plausible that major changes like the abolition of slavery or women's suffrage, had they not occurred when they did, would have happened a hundred years later than that they would *never* have happened. For now I'll focus on expected contingency on the order of hundreds of years. Values changes with this level of contingency are important in their own right, affecting many generations and often billions of people. But in the next chapter I'll also argue that there's a significant chance that the dominant values in the world over the next few centuries could get "locked in" and persist for an extremely long time. The values that are commonplace in the next few centuries might shape the entire course of the future.

To help us get clarity on the contingency of values over the course of history, we can consider an analogy to the contingency of biology over the course of evolution. Organisms have traits that affect their reproductive success, or "fitness." Evolution occurs because these traits vary, and some lead to more reproductive success than others.

Evolutionary contingency has been a topic of debate for decades. Evolutionary biologist Stephen Jay Gould thought that evolution is highly contingent. He claimed that if the "tape of life" were rerun, even very slight changes in the distant past could lead to huge differences in life on earth today.[56] Gould even speculated that the re-evolution of life with human-level intelligence would be unlikely.

The existence of evolutionary idiosyncrasies, like the elephant's trunk or the giraffe's neck, gives some evidence for contingency in evolution; if evolution were consistently convergent across a wide variety of environments, we would expect these traits to have evolved more than once.[57] Or consider New Zealand, which has been isolated ever since it split off from Australia about eighty million years ago. The island lacks any terrestrial native mammals, and in their absence, it became an "Empire of Birds," with birds evolving to occupy an unusual range of evolutionary niches.[58] These include the kiwi, which scavenges for insects on the forest floor; the kea, a parrot that, uniquely, lives in cold, high-altitude environments; and the now-extinct Haast's eagles, which are thought to have weighed up to fifteen kilograms, almost twice the size of any eagle alive today.[59]

However, in other cases we see convergent evolution, where species starting from very different places end up evolving the same traits. For example, insects, birds, pterosaurs, and bats all evolved the ability to fly despite different evolutionary histories. Similarly, we see streamlined bodies in fish, swimming mammals, and some molluscs. And crustaceans tend to evolve towards crab-like forms so often that the process of becoming a crab has its own name: carcinisation.[60]

Today, the consensus among biologists is that evolution can sometimes be contingent and sometimes noncontingent. This can be seen by considering what's called the "fitness landscape" (see Figure 3.2). In the fitness landscape, one or more dimensions measure the variation in an organism's traits; for example, for an elephant this could include its body mass, the length of its trunk, and its sociability. The final dimension measures that organism's evolutionary fitness as a function of its traits.[61]

The peaks in the landscape show which trait or combination of traits maximize the organism's fitness. Variation, like that caused by genetic mutation, causes individuals to occupy slightly different positions on the landscape. Those closer to the peak will be more likely to pass their traits on to the next generation. Sometimes there will be just one peak. Evolution will then push species towards that single peak no matter where on the landscape

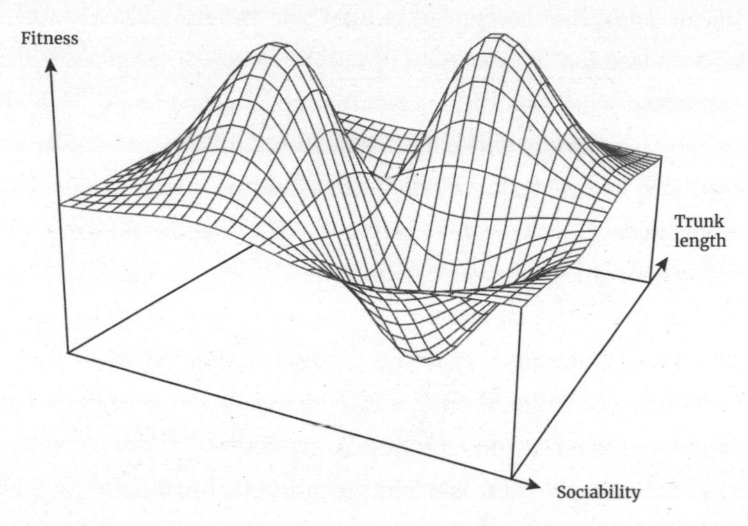

Figure 3.2. Simplified representation of a fitness landscape in biological evolution. It shows how an elephant's reproductive fitness might change depending on its sociability and trunk length. (For illustrative purposes only, not intended to make claims about actual elephants.)

they begin. For example, almost any sort of swimming animal will evolve a streamlined body.

In other cases, there are multiple peaks on the landscape, such as when there are different ways of adapting to the same environment. Beavers and platypuses both make slow-moving creeks and rivers their home, but they have very different traits. When there is more than one peak, we say there are *multiple equilibria*. This introduces contingency into evolution, since which peak an organism ends up climbing will depend on where it starts on the fitness landscape, how that landscape is shaped, and the randomness inherent to genetic mutation.

The contingency of biological evolution can be high if there are multiple equilibria. But even if there is only one equilibrium, *expected* contingency can be high if it simply takes a long time for that equilibrium to be reached—if evolution is slow at climbing the fitness landscape. For example, there were around seven hundred million years between the evolution of the first neurons and the evolution of human-level intelligence.[62] It's possible that human-level intelligence was always a peak on the fitness landscape, and it was just a very slow journey to get there. There could be many viable paths up to this peak, and if so, then the forms of intelligence that evolved would be contingent for seven hundred million years.

In recent decades, the theory of evolution and the fitness landscape has been used to understand the evolution of cultures, including values.[63] It can help us understand when and why values might be contingent.

In this theory, culture is understood broadly as any socially transmitted information, such as beliefs, knowledge, skills, and practices, though I will focus just on values. Cultural evolution can be described by the same three principles that govern Darwinian evolution:

- *variation*: cultural traits vary in their characteristics
- *differential fitness*: cultural traits with different characteristics have different rates of survival and reproduction
- *inheritance*: cultural traits can be transmitted from person to person via imitation or speech

So, for example, there are a variety of possible cultural attitudes to out-group members, from friendliness to hostility; some of these cultural attitudes will be better adapted to a given environment than others; those attitudes that are better adapted are more likely to be passed on to peers and to the next generation. In models of cultural evolution, one can get cultural competition between individuals and between groups.[64]

The lens of cultural evolution is helpful for understanding both the past and the future. As cultures interact with each other and adapt to their environment over time, new cultures and traits arise, and old cultures either evolve or are outcompeted. To be clear, I'm certainly not claiming that the traits which enable a culture to spread make it "better" than other cultures. We should be extremely worried that those cultures that have the highest fitness, and are most likely to win out over time, may not be those that are most desirable. As leading anthropologist Joe Henrich points out, norms that grossly devalue out-group members can be favoured by intergroup selection, motivating members of the tribe or nation to exterminate their competition.[65]

Just as there are fitness landscapes for organisms' traits, there are fitness landscapes for cultures' values. When such a landscape has a single peak, we should expect cultures to converge on the specific values represented by that peak—changing values would then be low in contingency. It doesn't seem surprising that norms in favour of caring for children are widespread: cultures without such norms are less likely to have healthy kids and less likely to thrive over time.[66] Similarly, cultures that seek to win converts and spread themselves as widely as possible, like proselytizing religions, seem more likely to grow than cultures that lack this trait. So, again, it doesn't seem surprising that many of the world's largest religions, like Christianity and Islam, value converting others to their faith.

However, there can also be multiple peaks on the fitness landscape, meaning that even in the long run different cultures could stably end up with very different values. For example, consider the phenomenon of conspicuous consumption: wealthy individuals buying goods to show off, in very public ways, how much wealth they have. The universality of conspicuous consumption suggests that there is cultural evolutionary pressure towards it. But the form that it takes is highly contingent: in some cultures, it can take the form of purchasing luxury goods; in others, it can take the form of philanthropy; in others still, it can take the form of owning enslaved people. Some of these forms of conspicuous consumption are far preferable to others.

For another example, note that in many religions it is important for adherents to demonstrate their piety or moral integrity. But different religions have developed very different ways of accomplishing this goal. Many Buddhists and Hindus demonstrate piety and moral integrity by being vegetarian; the same is not true for most Christians. This in part explains why one in five people in Asia say they are vegetarian while only one in twenty in Europe and North America do.[67] Similarly, China, Korea, and Vietnam all consume more than thirty kilograms of pork per person per year, whereas that number is close to zero for Muslim or Jewish countries, such as Iran, Pakistan, Indonesia, and Israel.[68] Religious norms around sex, marriage, work, and charity are similarly diverse; depending on one's religious background, the actions you take to show that you are an honourable or pious person can vary greatly. Though these different equilibria might be equally good from the perspective of cultural fitness, they can be much better or worse from a moral perspective. If you think that eating meat is morally wrong, then the fact that Hinduism and Buddhism converged on vegetarianism to show moral integrity is a very good thing.

A second reason for expecting multiple equilibria in moral attitudes is that value systems entrench themselves, suppressing ideological competition. To see this, consider some of history's many ideological purges. Between AD 1209 and 1229, the inappropriately named Pope Innocent III carried out the Albigensian Crusade with the goal of eradicating Catharism, an unorthodox Christian sect, in southern France. His goal was eventually accomplished: about two hundred thousand Cathars were killed in the Crusade, and Catharism was wiped out across Europe by 1350.[69] British history is also replete with examples of monarchs trying to suppress religious opposition: in the sixteenth century, Mary I had Protestants burned at the stake and ordered everyone to attend Catholic Mass; just a few years later, Elizabeth I executed scores of Catholics and passed the baldly named Act of Uniformity, which outlawed Catholic Mass and penalised people for not attending Anglican services.[70]

Ideological purges have been common through the twentieth century, too. On the Night of the Long Knives, Hitler crushed opposition from within his own party, cementing his position as supreme ruler of Germany. In Stalin's Great Terror, around one million people were murdered between 1936 and 1938,[71] purging the Communist Party and civil society of any opposition to him. In 1975–1976, Pol Pot seized power in Cambodia and turned it into a one-party state. Intellectuals were regarded as ideological enemies and could be murdered on the basis of the most meagre evidence; one refugee commented that you could be killed just for wearing eyeglasses.[72] In 1978, after consolidating his power, Pol Pot reportedly told members of his party that their slogan should be "Purify the Party! Purify the army! Purify the cadres!"[73] In a little more than three years, the Khmer Rouge killed about 25 percent of the Cambodian population.[74]

Entrenchment of values creates multiple equilibria because there is a significant element of chance in which value system becomes most powerful at a particular place and time, and because, once a value system has become sufficiently powerful, it can stay that way by suppressing the competition. Moreover, the theory of cultural evolution helps to explain *why* the predominant cultures in society tend to entrench themselves. Simply: those cultures that do not entrench themselves in this way are, over time, more likely to die off than those that do.

The final reason why the expected contingency of moral change can be high is that, even in cases where there is a single equilibrium, the process of reaching it might be slow. If selection pressures are not particularly strong or there are few opportunities for change, then cultures might find themselves at many different points on the fitness landscape and only converge at a peak after long periods of time. North Korea's governance culture seems much less fit than South Korea's, as evidenced by the former's decades-long economic stagnation.[75] But the North's regime has managed to survive for over seventy years.

With these considerations in mind, we can see today a number of value differences both within and between countries where those differences seem highly contingent. Antiabortion attitudes are strongest, and the laws against abortion strictest, in the Catholic countries of Chile, the Dominican Republic, El Salvador, Nicaragua, Vatican City, and Malta.[76]

For women's workforce participation, though there's a weak U-shaped trend with respect to GDP per capita (with the poorest and richest countries more likely to have greater workforce participation), there is an enormous amount of variation across countries. Muslim-majority countries like Somalia, Afghanistan, Iraq, Egypt, and Saudi Arabia have particularly low levels of female labour force participation, though of course there are exceptions, such as Kazakhstan.

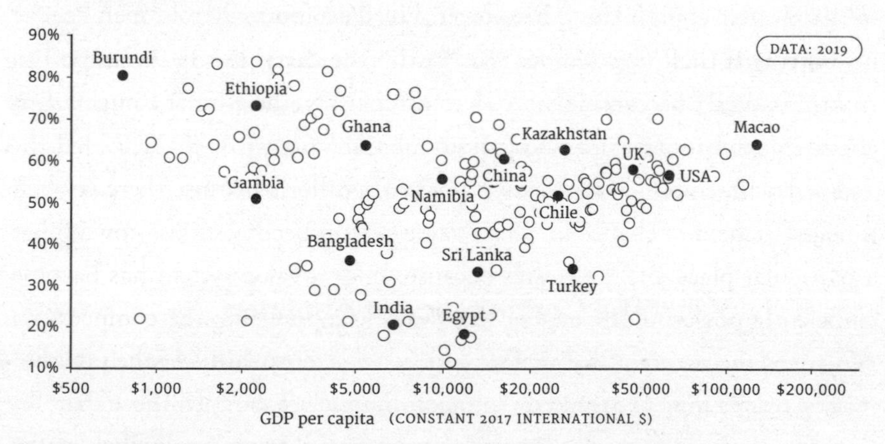

Figure 3.3. Proportion of women age fifteen and older who were economically active in 2019 against national per-capita income (adjusted for price differences between countries).

Women's workforce participation is reflected in cultural attitudes, too: Egypt and Peru both have a GDP per capita of about $12,000, but in Egypt about 80 percent of people think men have more right to a job than women do and fewer than 20 percent of women participate in the labour force, while in Peru only about 20 percent of people think men have more right to a job than women do and 70 percent of women participate in the labour force.[77]

Attitudes towards new biomedical technologies such as cloning and genetic enhancement vary substantially across countries, too. For example, the proportion of people who think it's acceptable to change a baby's genetic characteristics to make that child more intelligent ranges from 8 percent in Japan to 64 percent in India.[78] In general, countries in Asia seem more open to genetic enhancement than countries in Europe and the Americas, though there's a lot of regional variation.[79]

Similarly, across countries there are stark differences in willingness to fight for one's country (from 13 percent in Japan to 96 percent in Vietnam), in attitudes towards immigrants (in the average high-income country, 14 percent of the population is foreign-born, while just 2 percent of the populations of Japan and South Korea are), and in rates of vegetarianism (one study estimated that India has about ten times more vegetarians per capita than Brazil).[80] The same is true for levels of philanthropy: people in primarily Buddhist countries tend to give more to charity, with over 50 percent of people in Myanmar and Sri Lanka stating that they gave money to charity in the last month.[81] In many of these cases, facts about a country's history plausibly help explain the values its citizens have today.

Putting this all together, we have both theoretical reasons for expecting values to often be contingent and a number of examples where this contingency seems clear. But what about the example with which we led this chapter—the abolition of slavery? Might even that have been a contingent event?

The Contingency of Abolition

Slavery is so abhorrent that, before getting to grips with the historical scholarship on the topic, I assumed that abolition must have been inevitable. But now I'm not at all sure. Though it's impossible to know for certain, it's entirely plausible to me that, were the tape of history rerun a hundred times with slightly different starting conditions, in a significant proportion of those reruns, there would still be legal slavery in many or most countries in the world, even at today's level of technological development.[82]

The key question I'll look at is whether slavery's abolition was primarily the result of economic changes or changes in moral attitudes (though, of course, both were relevant). People often think that slavery's abolition was primarily an economic matter: Europe and its colonies were industrialising, which made slavery progressively less profitable; its abolition was just putting an end to an already-dying institution. This idea ultimately stems from the 1944 book *Capitalism and Slavery* by Eric Williams, an impressive scholar who later became the first prime minister of Trinidad and Tobago.

Williams's argument was a hugely important contribution, but it doesn't hold up to scrutiny, as demonstrated most convincingly by historian Seymour Drescher in his 1977 book *Econocide*. As Christopher Leslie Brown commented, "Since the publication of *Econocide*, few historians have continued to adhere to the economic interpretation of British abolition."[83] In correspondence, leading historians of abolition Manisha Sinha, Adam Hochschild, Michael Taylor, David Richardson, and Seymour Drescher himself said they broadly agreed with this claim.[84]

There are a few reasons for this. First, at the time of abolition slavery was enormously profitable for the British. In the years leading up to abolition, British colonies produced more sugar than the rest of the world combined, and Britain consumed the most sugar of any country.[85] When slavery was abolished, the shelf price of sugar increased by about 50 percent, costing the British public £21 million over seven years—about 5 percent of British expenditure at the time.[86] Indeed, the slave trade was booming rather than declining: even though Britain had abolished its slave trade in 1807, more Africans were taken in the transatlantic slave trade between 1821 and 1830 than in any other decade except the 1780s.[87] The British government paid off British slave owners in order to pass the 1833 Slavery Abolition Act, which gradually freed the enslaved across most of the British Empire.[88] This cost the British government £20 million, amounting to 40 percent of the Treasury's annual expenditure at the time.[89] To finance the payments, the British government took out a £15 million loan, which was not fully paid back until 2015.

The economic interpretation of abolition also struggles to explain the activist approach that Britain took to the slave trade after 1807. Britain made treaties, and sometimes bribes, to pressure other European powers to end their involvement in the trade and used the Royal Navy's West African Squadron to enforce those treaties.[90] Britain had some economic incentive here to prevent their rivals from selling slave-produced goods at lower prices than they could. But the scale of their activism doesn't seem worth it: from 1807 to 1867, enforcing abolition cost Britain almost 2 percent of its annual national income, several times what Britain spends today on foreign aid; political scientists Robert Pape and Chaim Kaufman described this campaign as "the most expensive international moral effort in modern history."[91] If the economic interpretation were correct, such activity would have been unnecessary because the slave trade would have been on its way out anyway.[92]

But might economic changes have made the end of slavery inevitable, at some later date, even if they were not the reason why the British Parliament abolished the slave trade? One could argue that as economies become increasingly mechanised, the value of slave labour decreases: the kinds of jobs which enslaved people were typically given—unpleasant work with easily measurable outputs—also seem like the kinds of jobs that are most likely to be automated.

This could give us some reason to think that the global proportion of enslaved people would have decreased over time, but it doesn't give us reason for thinking that slavery would have been entirely abolished. First, an enormous amount of labour is still unpleasant, low-skilled, and unmechanised, from fruit picking in the United States to mining and farming in lower-income countries. Sugarcane and cotton cultivation especially were very slow to be mechanised, even after US emancipation; mechanised harvesting became widespread in the South only after World War II.[93] Second, historically, many enslaved people were in roles not threatened by industrialisation, such as sex slaves and domestic servants. Finally, enslaved people have historically been employed in difficult-to-monitor work. In ancient Greece, for example, enslaved people often worked in skilled trades like metalworking and carpentry, in the civil service, in banking, and even in management positions in workshops or on large estates.[94]

Taking this evidence all together, we should conclude that slavery's end was not the inevitable result of economic factors; rather, it came about, in significant part, because of changing moral attitudes. Given this, we can ask how contingent it was for those changes in moral attitudes, and their enshrinement into law, to occur. This is difficult to ascertain because abolition essentially happened only once, in a single wave that swept the globe; we don't have access to independent historical experiments to see how things might have turned out. Is there just a single peak on the cultural fitness landscape, or are there many? Is the abolition of slavery more like the use of electricity—a more or less inevitable development once the idea was there? Or is it more like the wearing of neckties: a cultural contingency that became nearly universal globally but which could quite easily have been different?[95]

The optimistic view is that the moral changes that brought about slavery's end were more or less inevitable, part of the onward march of moral progress.[96] But it's hard to give strong support for this view. In particular, even if you think that the arc of the moral universe bends towards justice, that arc might still be very long. Perhaps in reruns of history, it takes a very long time at our current level of technological development for slavery to be abolished. If so, we might expect abolition to be contingent on the scale of centuries or even millennia.

Indeed, the history of the twentieth century, especially the rise of Nazism and Stalinism, shows how easy it is for moral regress to occur, including on the issue of free labour. During the Second World War, Nazi Germany used about eleven million forced labourers, 75 percent of whom were civilians; at its peak, forced labour accounted for about 25 percent of the country's workforce.[97] Similarly, the USSR under Stalin made widespread use of forced labour in gulag camps between 1930 and the 1950s, peaking at six million people, or 8 percent of the working population, in 1946.[98]

You might think that the progressive trend towards free labour in northwestern Europe supports the "march of moral progress" view and that the regresses in Nazi Germany and the USSR under Stalin were just blips. Slavery had died out in France and England by the end of the twelfth century, replaced by serfdom.[99] Serfs generally had more freedoms than enslaved people, and they typically could not be bought or sold, though they and their children were bound to a particular plot of land which they could not leave, and they were required to work for the land's owner.[100] Following the Black Death in the fourteenth century, serfdom was soon replaced by free labour throughout Western Europe.[101] Abolition might seem, therefore, to be the inevitable next step of this progressive trend.

However, the full historical picture is much more complicated. One enormous complication is the transatlantic slave trade itself: despite the domestic trend towards free labour, the European powers enslaved people on a massive scale; this alone makes the claim about a morally driven trend unclear at best. Second, we see no similar trend in other parts of the world.[102] In parts of Eastern Europe, serfdom intensified after the Black Death rather than declined.[103] In China, slavery waxed and waned over time. Slavery may have existed during the ancient Shang dynasty, which was founded before 1500 BC, and there is clear evidence of slavery during the Han dynasty (202 BC–AD 220).[104] De facto slavery continued in China in one form or another until the twentieth century. Several leaders attempted to reform or abolish slavery, often as part of political power struggles, but slavery repeatedly resurged when new dynasties came to power.[105] In the Liaodong province in 1626, for example, it was estimated that fully one-third of the population was enslaved by the Qing, and after the Manchu invasion and establishment of the Qing dynasty in 1636, slavery resurged for a time in other areas of China as well.[106] Slavery in China was abolished for good only in 1909.[107] Globally, it's hard to see abolitionism as part of even a stuttering historical trend towards moral progress on forced labour.

A more moderate view does not rely on the idea of moral progress but suggests that abolition was at least made very likely by a general tide of thought towards liberalism and free-market ideology in northwestern Europe. This is a position held by historian David Eltis.[108] In this view, once the idea took hold that people had equal rights, including the right to noncoercion by the state, logical consistency put pressure in favour of antislavery and abolitionist sentiment.

The independent emergence of antislavery currents among different groups of liberal intellectuals would, in my view, be strong evidence for this position. And there were seeds of abolitionist sentiment in countries other than Britain in the late eighteenth century. The most notable example is France. Several French thinkers, including Condorcet and Montesquieu, denounced slavery, and the French government made a half-hearted attempt to abolish it in 1794.[109] However, while abolitionist *sentiment* had emerged in France, the *campaign* to make it a legal reality grew out of British abolition. In fact, Jacques Pierre Brissot, founder of France's abolitionist group the Société des Amis des Noirs, was directly inspired by visiting London and meeting Thomas Clarkson.[110] Furthermore, the abolition law was repealed by Napoleon just eight years later, and France only abolished slavery permanently in 1848.[111]

It is also undoubtedly true that abolitionist sentiment was part of a wider package of more liberal thought, and a view that championed individual liberty yet endorsed slave owning should be, and often was, regarded as deeply morally inconsistent.[112] But we shouldn't think it obvious that liberal thought would lead to abolition. As historian Manisha Sinha has noted, "The heritage of the Enlightenment was a mixed blessing for Africans, giving a powerful impetus to antislavery but also containing elements that justified their enslavement. . . . No 'contagion of liberty' flowed inexorably according to its own logic to slaves."[113] The key question is how long inconsistencies in a moral worldview can persist.

Though logical inconsistency does seem to exert some pressure to change by giving advocates stronger arguments in favour of their views, there are many ways in which modern moral views have tolerated inconsistency for long periods of time. For example, tobacco and alcohol are legal and more or less socially acceptable in most countries around the world, whereas other drugs are illegal and their use is stigmatised. The abuse of dogs and cats can spark public outrage, while every year billions of animals suffer and are killed in factory farms.[114] Corporal punishment is considered a human rights violation, but ask yourself whether you would prefer to spend several years of your life behind bars or be flogged.[115] I'm not claiming that any of these are genuine moral inconsistencies: in each case you can give explanations to dissolve the seeming tension between these views and practices. But it certainly seems like our moral views host at least some deep inconsistencies, and that these inconsistencies can be remarkably persistent.

Crucially, these moral inconsistencies concern forced labour, too. Some forms of forced labour have persisted and sat more or less comfortably alongside liberalism. One example is conscription, which was used as late as the

1970s by the United States to force almost two million men to risk their lives in the war in Vietnam.[116] Another is penal labour. Consider, for example, the Mississippi State Penitentiary, better known as Parchman Farm. Beginning in 1901, the then governor of Mississippi, James K. Vardaman, ordered the building of a new prison that would operate as a profitable institution for the state. The result resembled "an antebellum plantation in every way, except that convicts replaced slave laborers."[117] The state government purchased nearly twenty thousand acres of land, racially segregated the inmates, and set them to work farming or picking cotton, often in intense heat and under threat of being whipped.[118] The penitentiary was highly profitable, making $26 million in today's money over 1912 and 1913.[119] These horrors might seem distant to us now. But Parchman stopped its most egregious practices only in the 1970s, and only under legal pressure.[120] And even today, thousands of prisoners in the United States work for the meagre wage of about one dollar per hour.[121] In some cases, they are not compensated at all. This is legal because the Thirteenth Amendment to the US Constitution abolished slavery and banned involuntary servitude, "except as a punishment for crime."[122]

Taking the possibility of such long-lasting inconsistency seriously, you might think that, were it not for the particular abolitionist campaign that did occur, then slavery might well have persisted even to this day. If so, then slavery's abolition was highly contingent. This is the view of Christopher Leslie Brown. In his book *Moral Capital*, he claims that "antislavery organizing was odd rather than inevitable, a peculiar institution rather than the inevitable outcome of moral and cultural progress. . . . In key respects the British antislavery movement was a historical accident, a contingent event that just as easily might never have occurred."[123]

Given how striking a view this is, there's more going for it than you might think. The key point is that the abolition movement was helped by many surprising or contingent factors. Brown emphasises the US War of Independence in particular. If the United States had instead remained part of the British Empire, Britain might have been more reluctant to jeopardise its uneasy relationship with the United States by taking a divisive action like abolishing the slave trade.[124] The plantation lobby would also have been bigger in a still-united empire. Finally, Brown notes that abolitionists in France struggled because they lacked the opportunities and status of those in England. Because abolitionist thought grew in France around the same time as the French and Haitian revolutions, abolitionist thought, Brown argues, became linked with violence and strife.[125]

According to Brown, in early nineteenth-century Britain, abolitionist action became a way to demonstrate virtue; in France, it did not. In this view, the abolitionist campaign occurred at a moment of plasticity, with multiple moral equilibria. Had things gone a different way over the course of a few crucial decades, antiabolition sentiment could have prevailed and then been further maintained by the plantation lobby.[126]

Moreover, even once the slave trade was abolished, the abolition of slavery itself was not a foregone conclusion. As historian Michael Taylor argues, British emancipation in 1833 could well have taken many decades longer to achieve than it did: "The ensuing, belated campaign for slave emancipation was no mere coda to the campaign against the slave trade. . . . There was absolutely nothing inevitable about its success."[127] Contingent events that helped the campaign for emancipation included parliamentary reforms in 1829 and 1832 that led to a largely abolitionist Parliament and the Jamaican Christmas Rebellion of 1831–1832, which brought more attention to colonial slavery and helped convince members of Parliament that slavery posed a threat to the British colonies.[128] Taylor also notes that two of the most important campaigners for emancipation, William Wilberforce and Zachary Macaulay, died between 1833 and 1838. If emancipation had not been achieved by 1838, he suggests, it could therefore have stalled altogether.[129] The difficulty of achieving emancipation was appreciated by campaigners at the time: in 1824, leading abolitionist Fowell Buxton reportedly would have been satisfied if slavery had been abolished within the next seventy years.[130]

Finally, even after Britain's abolition of slavery, it seems non-inevitable that emancipation would be achieved globally. Despite Britain's activist efforts, and despite the dominance of liberal ideas, global abolition still took over a century. Even into the 1930s, an estimated 20 percent of the population of Ethiopia was enslaved.[131] Slavery there was abolished only in 1942.[132] Saudi Arabia and Yemen were even later, abolishing slavery only in 1962.[133] There were still thousands of enslaved people in Saudi Arabia at the time.[134] Mauritania abolished slavery only in 1980 and only made owning people a criminal offense in 2007.[135] If there had been less effort to promote abolition globally, slavery could plausibly have persisted in some countries for even longer.

Putting this all together, we should be open to the striking idea that abolition was a contingent event. The view that abolition was more or less inevitable on economic grounds is not plausible. Regarding the question whether abolition was ultimately very likely, given the broader trend towards liberalism, or whether it was highly dependent on the success of the particular abolitionist campaign that was run, both answers have merit. On the latter view, abolition was brought about by the actions of a remarkably small number of people; on the former, it was the collective output of the many thousands who pushed French and British policy makers in the direction of a worldview that made slavery unacceptable. But either way, it was the actions of thinkers, writers, politicians, formerly enslaved activists, and enslaved rebels who together brought about the end of slavery. On either of these views, abolition was not preordained, and had history gone differently, the modern world could be one with widespread, legally permitted slavery.

What to Do

Once we take the contingency of moral norms seriously, we can start to consider a dizzying variety of ways in which the moral beliefs of the world could have been very different. Imagine if the Industrial Revolution had occurred in vegetarian-friendly India. Perhaps then the enormous rise of factory farming over the last century would never have occurred; the people in that alternative world would consider the suffering and death of tens of billions of animals every year in our world as an utter abomination.

Or imagine if Nazism had not grown in popularity. In the late nineteenth and early twentieth centuries, eugenics was widely supported among intellectuals in liberal countries like the United States, Britain, and Sweden.[136] If Nazism had not created such a strong opposition between eugenics and liberal ideas, then, horrifically, perhaps forced sterilisation and forced abortions would be widespread practices today. Or note that most cultures historically have been extremely patriarchal. If Roman attitudes towards gender had persisted in Western Europe, then perhaps the feminist movement could never have gotten off the ground.

I'm not claiming that we know the truth of any of these counterfactuals; it's impossible to know anything like this for certain. But given the theoretical reasons to expect multiple moral equilibria and the plausible examples of moral contingency that we can see today, we should not be confident that these very different moral worldviews couldn't have become widespread or even globally dominant. Certainly, the *expected* contingency of moral norms is high enough that the value of ensuring that the world is on the right track, morally, is enormously high. But if we take value changes seriously, which values should we promote, and how?

A longterm perspective favours value changes which are more generally applicable. For example, early Christian morality promoted both particular moral rules, like a prohibition against divorce, and general principles like the Golden Rule, that you should treat others as you would like to be treated. Particular moral rules can easily fail to achieve their intended purpose in contexts different to those in which they were originally proposed. The teachings of Jesus, though far from being feminist, were somewhat more progressive in terms of attitudes towards women than the extremely patriarchal societies of the time. This is especially because they banned divorce, which at the time was typically harmful to women because it was used by their families as a tool to make (or break) family alliances.[137] However, this is not true across all times and places; in the twentieth century, the legalisation of divorce was regarded as a major feminist victory. In contrast, the Golden Rule, if true at all, is true across all times and places. Promotion of that principle would stay relevant and, if true, have robustly positive effects into the indefinite future. Indeed, we saw it being used to further moral progress over 1,700 years after its Christian promotion, via the Quakers' recognition that the Golden Rule was inconsistent with the owning and trading of people.

This suggests that, as longtermists, when trying to improve society's values, we should focus on promoting more abstract or general moral principles or, when promoting particular moral actions, tie them into a more general worldview. This helps ensure that these moral changes stay relevant and robustly positive into the future.

The abolitionists demonstrate the importance of making moral change, but we can look to them as inspiration for *how* to make moral change, too.

Earlier, I mentioned that in the late eighteenth century, abolitionist Quakers would keep a print of Benjamin Lay in their house as a source of continued moral inspiration. I have followed their lead; a print of Lay sits next to my monitor, and he watches me as I write this book.

Lay was the paradigm of a moral entrepreneur: someone who thought deeply about morality, took it very seriously, was utterly willing to act in accordance with his convictions, and was regarded as an eccentric, a weirdo, for that reason. We should aspire to be weirdos like him. Others may mock you for being concerned about people who live on the other side of the planet, or about pigs and chickens, or about people who will be born in thousands of years' time. But many at the time mocked the abolitionists. We are very far from creating the perfect society, and until then, in order to drive forward moral progress, we need morally motivated heretics who are able to endure ridicule from those who wish to preserve the status quo.

To be clear, having "weird" beliefs does not mean engaging in weird actions. I think Benjamin Lay's guerrilla theatre was probably helpful in convincing the Philadelphia Quakers because they were already primed by their moral worldview to take antislavery sentiment seriously. But I suspect those same tactics would have backfired if used to try to convince the British public. For this next step of the campaign, activists like Anthony Benezet, who were able to repackage the Quakers' antislavery sentiment for a broader audience, were vital. US Founding Father Benjamin Rush wrote biographies of both Lay and Benezet. After describing Benezet as meek and gentle, Rush commented that he "completed what Mr Lay began."[138]

One social movement I'm particularly familiar with is the animal welfare movement, and through that I've seen the power of the combination of revolutionary beliefs and cooperative behaviour. For example, Leah Garcés is the president of Mercy for Animals. She has led Mercy for Animals to extraordinary success in recent years by joining other activist groups in convincing more than fifty US retailers and fast-food chains—including some of the biggest in the country, such as Walmart—to end their reliance on eggs from caged hens, reducing the suffering of tens of millions of animals each year.[139] The key to her success has been to treat her adversaries as human beings and find common ground with them. "The eventual goal should always be to sit down and negotiate with the so-called enemy and build solutions together," she told me. "Direct action and campaigns are important tactics for drawing attention to issues. . . . But they should be designed to lead to conversations, collaboration, and negotiations, not destruction of the enemy." Revolutionary beliefs; cooperative behaviour.[140]

If we succeed at improving the moral norms that society holds today, how long might that impact last? The history of religious and moral movements suggests that the impact could persist for centuries or even thousands of years. But could our impact last even longer than that? Might it even be that, at some point in the next few centuries, the values that guide the world could get locked in and continue to shape the future indefinitely? I'll turn to this idea in the next chapter.

CHAPTER 4

Value Lock-In

The Hundred Schools of Thought

In the sixth century BC in China, the collapse of the Zhou dynasty brought about a long period of conflict now known as the Warring States era. But this collapse also led to a vibrant era of philosophical and cultural experimentation—a golden age of Chinese philosophy later known as the Hundred Schools of Thought.[1]

During the Hundred Schools of Thought, philosophers would travel from state to state, developing their ideas and trying to persuade the political elite of their theories, moral commitments, and policy proposals.[2] Of the "hundred" schools, there were four leading philosophies.[3] Best known to us now is the philosophy of Kǒng Fūzǐ, or "Master Kǒng," better known in the West as Confucius. Confucians focused on promoting self-cultivation and moral refinement. They thought that, if you made a lifelong commitment to self-improvement, you could transform spiritually into a sage.[4] They likened cultivating your character to craftsmanship: cutting bone, carving a piece of horn, or polishing a piece of jade.[5]

Among other things, spiritual nobility involved the mastery of a range of social norms and cultural rituals advocated by the Confucians, as well as the careful refinement of your emotions.[6] Confucians encouraged obedience to authority, respect for your parents, and partiality to your family, rulers, and state. Rather than punishing wrong actions, Confucian legal principles punished wrong relationships: a son beating a father was a serious crime; a father beating a son was not.

A second school we now call Legalism.[7] Somewhat similar to Machiavellianism, Legalism took a dim view of human nature, regarding people as innately wicked and selfish. It emphasised the necessity of heavy punishments to prevent wrongdoing and the political importance of a wealthy government and a powerful military.

Third, there were the anti-authoritarian ideas expressed in the *Daodejing* and the *Zhuangzi* that later scholars referred to as Daoism. These books have traditionally been attributed to Lǎozǐ ("Old Master") and Zhuāngzǐ ("Master Zhuang"), respectively. Daoists believed that the Confucian attempt to control the world by promoting a rigid and unchanging set of social norms was foolhardy. They instead advocated spontaneous, noncoercive action that anticipates and responds to the ebb and flow of the world.[8]

Finally, there were the Mohists: followers of the fifth century BC philosopher Mòzǐ, or "Master Mò." Even though they are little known today, they were the main rival of the Confucians. They were so influential that their Confucian contemporary Mengzi said their teachings seemed to "fill the world."[9]

The Mohists argued that we should care about others just as much as we care for ourselves and that we should pursue whatever policies will produce the most benefit for all people.[10] They were the first consequentialists, endorsing the view that we should take whatever actions produce the best outcomes. Their philosophy has many similarities to that of the British utilitarians John Stuart Mill and Jeremy Bentham; the Mohists just got there two thousand years earlier.

Putting their radical ideas into practice, they argued that, to avoid wasting resources, people shouldn't own luxuries or consume too much.[11] They condemned the widespread nepotism of the time and advocated meritocracy instead. Being particularly distressed by war, some Mohists formed paramilitary groups devoted to protecting weaker cities. One commentator likened them to Jedi knights.[12]

There were bitter rivalries and intense criticism between these different schools. The Confucian philosopher Xúnzǐ wrote, "If your method is to follow Mòzǐ . . . then you may wander across the whole world, and even if you reach every corner of it, no one will not consider you base."[13]

The Hundred Schools of Thought ended in 221 BC, when the Legalism-influenced Qin conquered all of China and tried to purge any dissent from the new orthodoxy.[14] The emperor ordered the burning of unapproved books and prohibited all "private learning."[15] Disobedience was punished with death, and over four hundred dissenting scholars were murdered.[16] Legalism seemed to have won the war of ideas; Confucianism survived, but its influence was modest.[17]

The first Qin emperor was obsessed with the endurance of his rule. He declared that his empire would last for ten thousand generations, took advice from magicians who claimed they could create elixirs of immortality, and funded expeditions in search of mythical immortal beings.[18] His search was in vain, and he died in 210 BC at the age of forty-nine.

Popular revolt broke out after the emperor's death, and after years of conflict between competing factions, the Han general Liu Bang became the founding emperor of the Han dynasty.[19] The "ten thousand–generation" Qin Empire lasted just fifteen years.

By now, Legalism had been tainted by its association with the Qin and its oppressive policies. During the first years of the Han, imperial decisions were informed by a blend of Legalism, Confucianism, and Daoism.[20] Confucianism had no special status initially,[21] but a combination of luck and skilful politicking meant that Confucianism soon emerged as the orthodox ideology of the Chinese Empire. Emperor Xuan, who reigned from 74 to 48 BC, made Han dynasty China the first Confucian empire.[22]

Of course, the Confucians still had to contend with competitors. After the fall of the Western Han dynasty, Buddhism spread throughout China, and for much of the relatively open Tang dynasty of AD 618–907, Confucianism, Daoism, and Buddhism were all popular and tolerated by the state.[23] But starting in the mid-ninth century, Confucianism once again emerged as China's dominant public ideology.[24] For over a thousand years, every educated person in China was required to master the Confucian canon, and for seven hundred of those years, basic literacy was taught via the *San Zi Jing*, a Confucian classic written especially for children.[25]

Today, more than 2,500 years after Confucius's death, Confucianism's influence in China has waned.[26] It lost its position as official state philosophy in 1912, when it became fashionable to see Confucianism as an obstacle to China's economic development. But the influence of Confucianism on the history of China and other "Confucian heritage" countries is undeniable.

Even today, people from Confucian-heritage countries have distinctively Confucian views on what they think is important in life, how they expect their children to behave, and what their hopes are for the future.[27] But if events had unfolded differently two thousand years ago, plausibly instead it could have been Legalism, Daoism, Mohism, or some blend of these that ruled China for two thousand years.

The Persistence of Values

Values can be highly persistent.[28] A familiar but remarkable fact is that the best-selling book this year, as every year, is the Bible,[29] completed almost two thousand years ago. The second best-selling book is the Quran.[30] Confucius's *Analects* still sells hundreds of thousands of copies annually.[31] Every day, quotes from these sources influence political decision-making around the world.

The Babylonian Talmud, compiled over a millennium ago, states that "the embryo is considered to be mere water until the fortieth day"—and today Jews tend to have much more liberal attitudes towards stem cell research than Catholics, who object to this use of embryos because they believe life begins at conception.[32] Similarly, centuries-old dietary restrictions are still widely followed, as evidenced by India's unusually high rate of vegetarianism, a $20 billion kosher food market,[33] and many Muslims' abstinence from alcohol.

In this chapter I discuss *value lock-in*: an event that causes a single value system, or set of value systems, to persist for an extremely long time. Value lock-in would end or severely curtail the moral diversity and upheaval that we are used to. If value lock-in occurred globally, then how well or poorly the future goes would be determined in significant part by the nature of those locked-in values. Some changes in values might still occur, but the broad moral contours of society would have been set, and the world would enact one of only a small number of futures compared to all those that were possible.[34]

The rise of Confucianism illustrates the phenomenon of lock-in. The Qin tried and failed to lock in Legalism; the Han succeeded in locking in Confucianism for over a thousand years. But the lock-in that could occur this century or the next might last much longer—even indefinitely.

This sounds extreme, and as a warning, this chapter will discuss some ideas that will seem weird or sci-fi. But technology is changing rapidly, and technological advances could radically alter the dynamic of moral change that we are used to. When taking the interests of future generations seriously, we simply cannot dismiss major technological advances out of hand. Consider how someone in 1600 would react to the idea that, within two dozen generations, we would be able to make light and fire with the flick of a switch, and would do so dozens of times a day, without a second thought. Or that we could see anyone, anywhere in the world, immediately, in real time, on a device we carried in our pocket. Or that we could fly in the skies, or walk on a celestial body. We simply know that, given continued technological progress, there will be major change over the coming centuries.

Previous technology has already enabled values to persist for longer, and with higher fidelity, than they could otherwise have done. Writing, for example, was crucial, enabling complex ideas to be transmitted many generations into the future without inevitable distortion by the failures of human memory. The persistence of religious values, or moral worldviews like Confucianism, would not have been possible without writing as a technology.

In Chapter 2 I described the phenomenon of "early plasticity, later rigidity": that it can be much easier to influence the norms, standards, and laws surrounding a technology, idea, or country when they are still new than later on, when things have settled. In China, the Hundred Schools of Thought was a period of plasticity. Like still-molten glass, during this time the philosophical culture of China could be blown into one of many shapes. By the time of the Song dynasty, the culture was more rigid; the glass had cooled and set. It was still possible for ideological change to occur, but it was much more difficult than before.

We are now living through the global equivalent of the Hundred Schools of Thought. Different moral worldviews are competing, and no single worldview has yet won out; it's possible to alter and influence which ideas have prominence. But technological advances could cause this long period of diversity and change to come to an end.

When thinking about lock-in, the key technology is artificial intelligence.[35] Writing gave ideas the power to influence society for thousands of years; artificial intelligence could give them influence that lasts millions. I'll discuss *when* this might occur later; for now let's focus on why advanced artificial intelligence would be of such great longterm importance.

Artificial General Intelligence

Artificial intelligence (AI) is a branch of computer science that aims to design machines that can mimic or replicate human intelligence. Because of the success of machine learning as a paradigm, we've made enormous progress in AI over the last ten years. Machine learning is a method of creating useful algorithms that does not require explicitly programming them; instead, it relies on learning from data, such as images, the results of computer games, or patterns of mouse clicks.

One well-publicised breakthrough was DeepMind's AlphaGo in 2016, which beat eighteen-time international champion Go player Lee Sedol.[36] But AlphaGo is just a tiny sliver of all the impressive achievements that have come out of recent developments in machine learning. There have also been breakthroughs in generating and recognising speech, images, art, and music; in real-time strategy games like *StarCraft*; and in a wide variety of tasks associated with understanding and generating humanlike text.[37] You probably use artificial intelligence every day, for example in a Google search.[38] AI has also driven significant improvements in voice recognition, email text completion, and machine translation.[39]

The ultimate achievement of AI research would be to create *artificial general intelligence*, or AGI: a single system, or collection of systems working together, that is capable of learning as wide an array of tasks as human beings can and performing them to at least the same level as human beings.[40] Once we develop AGI, we will have created artificial *agents*—beings (not necessarily conscious) that are capable of forming plans and executing on them in just the way that human beings can. An AGI could learn not only to play board games but also to drive, to have conversations, to do mathematics, and countless other tasks.

So far, artificial intelligence has been narrow. AlphaGo is extraordinarily good at playing Go but is incapable of doing anything else.[41] But some of the leading AI labs, such as DeepMind and OpenAI, have the explicit goal of building AGI.[42] And there have been indications of progress, such as the performance of GPT-3, an AI language model which can perform a variety of tasks it was never explicitly trained to perform, such as translation or arithmetic.[43] AlphaZero, a successor to AlphaGo, taught itself how to play not only Go but also chess and shogi, ultimately achieving world-class performance.[44] About two years later, MuZero achieved the same feat despite initially not even knowing the rules of the game.[45]

The development of AGI would be of monumental longterm importance for two reasons. First, it might greatly speed up the rate of technological progress, economic growth, or both. These arguments date back over sixty years, to early computer science pioneer I. J. Good, who worked in Bletchley Park to break the German Enigma code during World War II, alongside Alan Turing and, as it happens, my grandmother, Daphne Crouch.[46]

Recently, the idea has been analysed by mainstream growth economists, including Nobel laureate William Nordhaus.[47] There are two ways in which AGI could accelerate growth. First, a country could grow the size of its economy indefinitely simply by producing more AI workers; the country's growth rate would then rise to the very fast rate at which we can build more AIs.[48] Analysing this scenario, Nordhaus found that, if the AI workers also improve in productivity over time because of continuing technological progress, then growth will accelerate without bound until we run into physical limits.[49]

The second consideration is that, via AGI, we could automate the process of technological innovation. We have already seen this recently to some extent: DeepMind's machine-learning system AlphaFold 2 made a huge leap towards solving the "protein folding problem"—that is, how to predict what shape a protein will take—reaching a level of performance that had been regarded as decades away.[50] If AGI could quite generally automate the process of innovation, the rate of technological progress we have seen to date would greatly increase. This acceleration would apply to the design of AI systems themselves, in a positive feedback loop. This idea was formalised in a model by some leading growth economists; again, they found that AI could produce extraordinarily fast—and accelerating—rates of growth.[51]

It's not inevitable that AI will impact technological progress in this way. Indeed, the authors of the models I've referenced emphasise that accelerating growth rates hold only under some conditions.[52] Perhaps, for example, there are some crucial inputs that are very hard to automate; perhaps these include the manufacturing of computer chips, or the mining of ores to create those chips, or the building of power plants to power the server farms the AI systems rely on. If so, then the slow growth in these areas would constrain the overall rate of progress.

However, given the clear mechanisms by which AI could generate far faster growth rates, we should take this possibility very seriously. Economies could double in size over months or years rather than decades.

This might seem implausible, but, remarkably, moving to much faster rates of economic growth would be a continuation of historical trends. We are used to thinking about growth in terms of a steady exponential, where a country's economy grows by a few percent every year. But over the long run growth rates have accelerated. In the early agricultural era, the global rate of economic growth was around 0.1 percent per year; nowadays, it is around 3 percent per year.[53] Before the Industrial Revolution, it took many centuries for the world economy to double in size; now it doubles every twenty-five years.

It's not clear how best to understand this. Perhaps history was a succession of distinct exponential "growth modes"—moving from a hunter-gather economic era to an agricultural era to an industrial era.[54] Or perhaps economic history is just a single faster-than-exponential but noisy trend, with rates of growth steadily accelerating over time. In this latter view, the last one hundred years of relatively stable growth rates are anomalously slow.[55] But in either the "growth modes" view or the "single faster-than-exponential trend" view, we should be open to the idea that growth rates might be much higher in the future than they are today. Given that growth rates have increased thirtyfold since the agricultural era, it's not crazy to think that they might increase tenfold again; but if they did, the world economy would double every two and a half years.[56]

An increase in the rate of technological progress is the first reason why AGI would be a monumental event. The second reason, crucial from a longterm perspective, is AGI's potential longevity.[57]

In Chapter 1 we saw that Shakespeare and Horace really might have achieved immortality through their poetry. Information can persist indefinitely because the cost to replicate it is so tiny. But software is just complex information. It can be replicated easily. For example, one of the first commercially available computer games was *Pong* by Atari, released in 1977.[58] You can still play it today online.[59] Though eventually all original Atari consoles will rust and crumble, *Pong* will live on. The software that defines *Pong* is replicable, and if every future generation is willing to pay the tiny cost of replicating this little piece of history, it will continue to persist. *Pong* could last as long as civilisation does.

There's nothing different in principle between the software that encodes *Pong* and the software that encodes an AGI. Since that software can be copied with high fidelity, an AGI can survive changes in the hardware instantiating it. AGI agents are potentially immortal.

AI and Entrenchment

These two features of AGI—potentially rapid technological progress and in-principle immortality—combine to make value lock-in a real possibility.

Using AGI, there are a number of ways that people could extend their values much farther into the future than ever before. First, people may be able to create AGI agents with goals closely aligned with their own which would act on their behalf. A lot of work has already been done on how to align AI with human intentions, such as by developing AI systems that are able to copy the behaviour of people or infer their goals. Second, the goals of an AGI could be hard-coded: someone could carefully specify what future they want to see and ensure that the AGI aims to achieve it. Third, people could potentially "upload": scan their brain at high resolution and then emulate its structure on a computer. Just as modern computers can enable you to play retro computer games by running an emulation of old video consoles, a future computer could replicate the functions of a human brain by emulating it digitally.[60] This emulation would be functionally the same as the uploaded mind, living on in digital form. Finally, some combination of these techniques could be used. The first two pathways are simply extensions of existing AI research.[61]

Would we wield such unprecedented power responsibly? Worryingly, the pursuit of value lock-in has been common throughout history. We saw that when the Qin took control of China, they undertook a programme to systematically eradicate competing schools of thought; similarly, the Han systematized Confucian teachings to the detriment of competing schools. The Mohists, too, desired to lock in their own values indefinitely, if only they had the power. They saw moral disagreement as the biggest problem in the world and thought that the solution was to ensure that everyone had the same values. They told a parable of bygone "sage kings" who set up a chain of command from themselves all the way down to the lowest peasants: at each step of the chain, the subordinate would copy the values of their superior perfectly; this would carry on until Mohist values had been perfectly transmitted to all members of society.[62]

Similarly, in the previous chapter I gave examples of religious crusades and ideological purges that aimed to eliminate people who advocated for different values. Some of these, like Stalin's Great Purge, were highly successful.[63] In the previous chapter I discussed how the theory of cultural evolution explains why many moral changes are contingent. The same theory also explains why they can be so persistent. When we look at history, we see that the predominant culture in a society tends to entrench itself, eliminate the competition, and take steps to replicate itself over time. Indeed, many moral views regard their own lock-in as desirable.[64] As I mentioned in the last chapter, cultural evolution partly explains why: those cultures that do not entrench themselves in this way will, over time, be more likely to die off than those that do. This results in a world increasingly dominated by cultures with traits that encourage and enable entrenchment, and thus persistence.[65]

The pursuit of lock-in could also be a side effect of the pursuit of immortality (for example, via mind uploading) combined with an unwillingness to give up power. A desire for immortality has been very common throughout history. As early as the second millennium BC, the *Epic of Gilgamesh* told a story in which Gilgamesh, who was probably a real-life king, attempts to secure eternal life.[66]

We also already noted the first Qin emperor's search for immortality. Here he was not unique; for thousands of years in China, immortality on earth was a popular aim.[67] One history of Chinese chemistry describes dozens of substances and potions for eternal life tested by emperors and their alchemists throughout much of this period.[68]

In the last century, many authoritarian or totalitarian rulers were interested in or actively pursued life extension.[69] Stalin expressed an interest in the topic, and according to one Soviet defector, this prompted scientists to make life extension "a central subject of Soviet medical research."[70] North Korea's Kim Il-sung set up a longevity centre devoted to keeping him alive and received blood transfusions from citizens in their twenties in an attempt to live longer.[71] Nursultan Nazarbayev, the authoritarian ruler of Kazakhstan between 1990 and 2019, tasked Kazakh scientists with "the prolongation of life." But after spending two years and millions of dollars, they disappointingly only managed to produce a probiotic yogurt called Nar.[72]

More recently, many wealthy techno-optimists have provided hundreds of millions of dollars in funding for biomedical R&D companies aiming to achieve indefinite life spans. Amazon CEO Jeff Bezos and PayPal cofounder Peter Thiel have both invested in San Francisco–based Unity Biotechnology, a company whose mission is to prevent aging.[73] In 2013, Google launched the company Calico, which also aims to combat aging, with more than a billion dollars in funding.[74] Ambrosia, a California start-up, charges its elderly customers $8,000 for injections of two and a half litres of blood plasma harvested from teenagers.[75]

Even if aging cannot be cured in our lifetime, some people plan to punt the problem to the future by paying for cryonics: having their body or severed head frozen in the hope that resurrection will be possible with future technology. Whole-body cryopreservation with the Alcor Life Extension Foundation costs $220,000; it costs less than half that if one merely preserves one's head.[76] Some entrepreneurs hope to abandon meat-based bodies altogether and live on in digital form through computer emulation of their brains. Nectome, a Y Combinator–funded start-up that preserves brains with the hope that future generations will scan and upload them, counts Silicon Valley entrepreneur Sam Altman as a customer. Nectome's founder, Robert McIntyre, describes the service as "100% fatal."[77]

If the aim of locking in values and the desire for immortality have been so common throughout history, then we should expect many people to have those aspirations in the future, too. AGI could allow them to become reality.

AGI could affect *who* has power, too. AGI might be developed by a company or a military, and power could be in their hands rather than the hands of states. International organisations or private actors may be able to leverage AGI to attain a level of power not seen since the days of the East India Company, which in effect ruled large areas of India in the eighteenth and nineteenth centuries. AGI could not just upend the international balance of power; it could also reshape which kinds of actors matter most in world affairs.

If we don't design our institutions to govern this transition well—preserving a plurality of values and the possibility of desirable moral progress—then a single set of values could emerge dominant. They may be those championed by a single individual, the elites of a political party, the populace of a country, or even the whole world.

If this happened, then the ruling ideology could in principle persist as long as civilisation does. AGI systems could replicate themselves as many times as they wanted, just as easily as we can replicate software today. They would be immortal, freed from the biological process of aging, able to create back-ups of themselves and copy themselves onto new machines whenever any piece of hardware wears out. And there would no longer be competing value systems that could dislodge the status quo.

This section so far has been premised on people aligning AGI with their goals. But they may well fail. The attempt to lock in values through AGI would run a grave risk of an irrecoverable loss of control to the AGI systems themselves, which, if misaligned and uncontrolled, would kill the AGI's developers as well as everyone else. This is the risk I now turn to.

AI Takeover

If we build AGI, it will likely not be long before AI systems far surpass human abilities across all domains, just as current AI systems far outperform humans at chess and Go. And this poses a major challenge. To borrow an analogy from Ajeya Cotra, a researcher at Open Philanthropy, think of a child who has just become the ruler of a country.[78] The child can't run the country themselves, so they need to appoint an adult to do so in their place. Their aim would be to find an adult who will act in accordance with their wishes. The challenge is for the child to do this—rather than, say, appointing a schemer who is good at deceitful salesmanship but once in power would pursue their own agenda—even though the adults are much smarter and more knowledgeable than the child is.

This risk was the focus of Nick Bostrom's book *Superintelligence*. The scenario most closely associated with that book is one in which a single AI agent designs better and better versions of itself, quickly developing abilities far greater than the abilities of all of humanity combined. Almost certainly, its aims would not be the same as humanity's aims. And in order to better achieve its aims, it would try to gain resources and try to prevent threats to its survival.[79] It would therefore be incentivised to take over the world and eliminate human beings or permanently suppress them.[80]

Recent work has looked at a broader range of scenarios.[81] The move from subhuman intelligence to superintelligence need not be ultrafast or discontinuous to pose a risk. And it need not be a single AI that takes over; it could be many. We could see human beings gradually lose control as AI systems become a larger and larger share of the world economy. Eventually we would share the fate of, say, chimpanzees or ants vis-à-vis humans: ignored at best and with no say over the future of civilisation. To avoid such disempowerment, people would need to ensure that the AIs did what their operators wanted them to do. This is known as the "alignment" problem.[82] It's discussed at length in other excellent books, like *Superintelligence*, Stuart Russell's *Human Compatible*, and Brian Christian's *The Alignment Problem*, so I won't go into it in depth here.

Often the risk of AI takeover is bundled with other risks of human extinction. But this is a mistake. First, not all AI takeover scenarios would result in human extinction. If human beings wanted to make chimpanzees extinct, we could—but we choose not to. We have no reason to, because they are not a threat to human hegemony. Similarly, even if superintelligent AGIs take over, they might well have so much more power than humans that they have no need to kill us off.

Second, and more important, from a moral perspective AI takeover looks very different from other extinction risks. If humanity were to go extinct from a pandemic, for example, and no other species were to evolve to build civilisation in our place, then civilisation would end, and life on earth would end when the expanding sun renders our planet uninhabitable. In contrast, in the AI takeover scenarios that have been discussed, the AI agents would continue civilisation, potentially for billions of years to come. It's an open question how good or bad such a civilisation would be.

As an analogy, imagine you are a member of an island nation considering two ways in which your nation might end. First, a plague could kill everyone on your island; the island would thereafter be uninhabited. Second, colonisers could invade, wipe out everyone on the island, and afterwards build a different nation on the island, with (let us assume) worse values.

The future of this island would be very different under each of these two scenarios, and its moral assessment would be very different, too. Even if you thought that the absence of future generations was not a moral loss, and even if you thought that the extinction of humans on your island via plague would be a good thing, morally speaking, you would still want to prevent the invasion of your island by the colonisers. By preventing the plague, you would enable the continuation of your nation rather than nothingness. By preventing colonisation, you would enable the continuation of your nation rather than its replacement by some other nation with worse values.

In the same way, even if superintelligent AGI were to kill us all, civilisation would not come to an end. Rather, society would continue in digital form, guided by the AGI's values. What's at stake when navigating the transition to a world with advanced AI, then, is not *whether* civilisation continues but *which* civilisation continues.[83]

For this reason, even if you think that the absence of future generations is not a moral loss or that the end of civilisation would be a good thing (issues that I discuss in Part IV), it's still very important to avoid AI takeover or the lock-in of bad values. There will be future generations of intelligent beings either way, and by preventing the takeover of the world by an AI with bad values, you are changing how good or bad the future is over the course of civilisation's life span. That's the main effect, rather than any impacts on civilisation's life expectancy.

The key issue, in my view, is not whether humans or AIs are in control; either way, AGI is a way for values to get locked in indefinitely. The key issue is which values will guide the future. Those values could be narrow-minded, parochial, and unreflective. Or they could be open-minded, ecumenical, and morally exploratory.

If lock-in is going to occur either way, we should push towards the latter. But transparently removing the risk of value lock-in altogether is even better. This has two benefits, both of which are extremely important from a longtermist perspective. We avoid the permanent entrenchment of flawed human values. And by assuring everyone that this outcome is off the table, we remove the pressure to get there first—thus preventing a race in which the contestants skimp on precautions against AGI takeover or resort to military force to stay ahead.

How Long Till AGI?

You might think that my discussion so far is idle speculation, because AGI is still thousands of years away. But this would be a mistake.

It's certainly true that we don't *know* when we will build AGI. But uncertainty cuts both ways. Technological developments can be surprisingly slow, but they can also be surprisingly fast. For example, the British Indian geneticist J. B. S. Haldane was one of the first people to grasp the sheer scale of the future and its moral import. In a 1927 essay called "The Last Judgment," he expresses a vision for the human future over the next forty million years.[84] To my knowledge, it is the first time that anyone predicted that humanity could spread across the galaxy. Yet in the same essay, Haldane predicted it would take over eight million years for us to make a return trip to the moon.[85]

In some cases, even when there is a clear trend in technological progress, people can fail to pick up on it. For instance, the cost of solar panels has been consistently declining on an exponential trend for more than forty years.[86] But all mainstream economic models have failed to extrapolate this trend forward and so have tended to be too pessimistic on solar deployment.[87] Exponential progress, let alone superexponential progress, is hard for us to grasp.

AGI might still be far in the future. But it might come soon—within the next fifty or even twenty years.

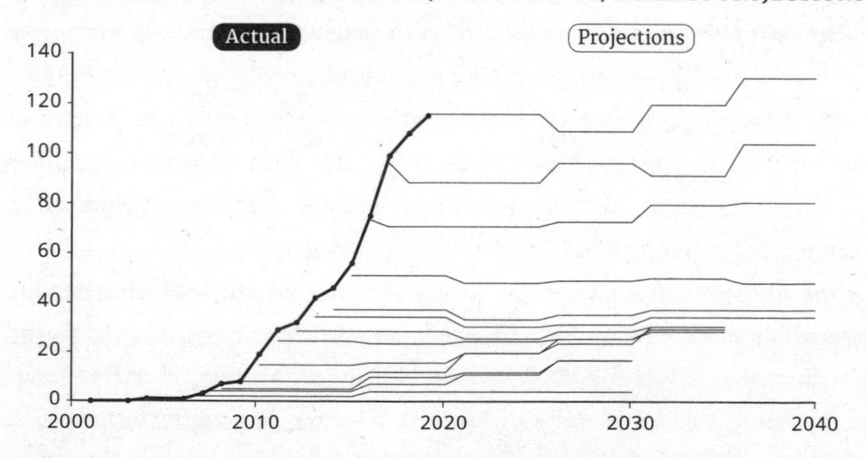

Figure 4.1. Global solar capacity has outpaced all projections by the International Energy Agency since 2006. Graph shows capacity growth per year (rather than cumulative total).

The most weighty evidence for this is marshalled by Ajeya Cotra. Her report forecasts trends in computing power over time and compares those trends to the computing power of the brains of biological creatures and the amount of learning they require to attain their abilities.[88] Using what we know from current neuroscience, today's AI systems are about as powerful as insect brains, and even the very largest models are less than 1 percent as powerful as human brains.[89] In the future, this will change.

The cost of computation is exponentially falling while both the efficiency of AI systems and the budgets of the largest machine-learning training runs are exponentially increasing.[90] Based on extrapolations of these trends and our best guesses from neuroscience, Cotra found that we are likely to train AI systems that use as much computation as a human brain within roughly the next decade, and that we may well have enough computing power to essentially simulate the complete history of biological evolution by the end of this century.[91]

These comparisons involve a lot of uncertainty, such as in how much computation the human brain uses. Taking this uncertainty into account, Cotra gives a greater than 10 percent chance of AGI by 2036 and a 50 percent chance of AGI by 2050.[92]

On a podcast discussing her work, Cotra says that, as a result of her research, she's now "thinking of AI much more viscerally, as this onrushing tide." She acknowledges it's "a quite extreme and stressful and scary conclusion, because I'm forecasting a date by which the world has been transformed."[93]

But isn't this timeline inconsistent with machine-learning experts' views on the matter? No. In 2016, Katja Grace, founder of the think tank AI Impacts, ran what's currently the most comprehensive survey.[94] About 350 top machine-learning researchers estimated by what year "unaided machines can accomplish every task better and more cheaply than human workers," a notion very similar to AGI.[95]

The main conclusion from the survey was that machine-learning experts as a whole don't have stable and consistent beliefs about the matter. The average response was that there is a 10 percent probability of unaided machines being able to accomplish every task better and more cheaply than human workers by 2025 and a 50 percent probability by 2061.[96] But when asked about a different operationalization of AGI—"when for any occupation, machines could be built to carry out the task better and more cheaply than human workers"—the average response was a 50 percent probability of AGI by 2138, more than twice as many years into the future as the previously quoted prediction of 2061.[97] Those surveyed also predicted it would take much longer for AI systems to outperform humans at AI research than to outperform humans at "every task," which is impossible.[98]

This means that we shouldn't place much weight on surveys of machine-learning experts when trying to predict timelines to AGI. But it also means that we cannot at all say that the experts think that AGI is centuries away: under some framings of the question at least, they say that AGI might well come within the next few decades.

A different response you might have is that we have been trying and failing to build AGI for decades, with overinflated hype along the way, so any future prediction should be treated with scepticism. But the previous hype is commonly exaggerated—there was widespread overoptimism in the 1950s and '60s, but there were also many dissenting voices.[99] And, more important, people crying wolf in the past doesn't tell us much about what we should think now. Another researcher at Open Philanthropy, Tom Davidson, created an estimate of timelines to AGI based only on how long we've been doing AI research, how much more research effort we should expect in the future, and analogues to comparable historical events. His best guess was that, if you only had access to this information, you should think that the probability of AGI by 2036 is around 8 percent. You should then adjust this estimate up or down based on additional information, like recent achievements in AI.[100]

All these sources of evidence are fallible. Long-run forecasting seems difficult enough to me that we should remain highly uncertain. But these threads, in combination with the astonishing progress that has been made in AI over the last decade, should make us take short timelines to AGI seriously. I don't think that one could reasonably go lower than a 10 percent chance of AGI in the next fifty years. But if so, there's a very significant chance that one of the most important developments in all of history will occur within our lifetimes.

Culture and Lock-In

What if AGI is centuries away? It would still be of enormous importance because it creates a date at which the predominant values of a time could get locked in—and what we do over the coming years could affect what values are predominant when AGI is first built. The examples of religions and other moral worldviews already show that values can persist for centuries, though they evolve along the way. But values could become even more persistent in the future if a single value system were to become globally dominant. If so, then the absence of conflict and competition would remove one reason for change in values over time.

Conquest is the most dramatic pathway by which a single value system can become globally dominant, and it may well be the most likely. In the next chapter I'll suggest that there's a significant chance of a third world war in our lifetimes. If that happens, perhaps the outcome will be a single world government and the global promotion of that government's ideology.

Indeed, cultural conquest is quite commonplace. When we look at the map of the distribution of world religions, much of it can be explained by the history of conquest and colonialism. Protestant Christianity is the most common religion in the United States because of British colonialism; Catholicism is the most common religion in Latin America because of Spanish and Portuguese colonialism.[101] Afghanistan was primarily Buddhist for around eight hundred years, from the second century BC to around AD 650.[102] The start and end of this period were both driven by conquest: first, the conquest by the Buddhist Mauryan Empire and some time later the Kushan Empire; second, the conquest by the Rashidun Caliphate, the first caliphate established after the death of the prophet Muhammad.[103] Today, almost 1,400 years after this conquest, 99.7 percent of the population of Afghanistan is Muslim.[104]

And there are historical examples of ideologies that have sought long-lasting global domination. This was true of the Nazis, who referred to their empire as the "Thousand-Year Reich." Similarly, the vision of global communism was promoted by the Soviet-controlled organization Comintern, which before World War II held seven World Congresses designed to further the aim of world revolution.[105]

But even if no single value system conquers all others, we might still converge to a single hybrid value system that is a blend of multiple sets of values, like different colours of paint mixed together to produce a new hue. This might look like convergence to a single "best" moral worldview, but really it's just a function of what values the world started with and how heavily represented each were.

The nature of the values that the world converges on would depend on how powerful different value systems were before that point. And this can be affected by many factors. Conquest that falls short of global domination is one. A second way for a culture to become more powerful is immigration. For example, for the last 130 years, the United States has been the world's largest economy.[106] By definition, the size of a country's economy is given by its GDP per person and its population size. And the United States' current population size is due, in part, to the high rates of immigration from Europe to North America from 1607 onwards and especially after 1830. In the future, countries that maintain high rates of immigration and cultural assimilation will grow in size and power; indeed, journalist Matt Yglesias recently proposed that, in order to maintain global influence, the United States should radically increase immigration, aiming to have a population of one billion people.[107]

A third way in which a cultural trait can gain influence is if it gives one group greater ability to survive or thrive in a novel environment than some other group. You might think that this consideration is not terribly important, because people already inhabit almost all the remotely habitable areas of Earth. But when we look to the future, there is a vast territory that civilisation might expand into: space. Even within our own solar system, the potential energy outside of Earth is over a billion times greater than that on Earth; even within our own galaxy, there is billions of times the energy outside of our solar system than within.[108] If one culture made greater efforts to settle in space or had greater ability to do so, then eventually it would dwarf any culture that chose to remain earthbound.

A final way in which one culture can outcompete another is via higher long-run population growth. For example, through a combination of high conversion rates and high fertility rates, Christianity rose to become the predominant religion in Europe over what was a remarkably short time period in historical terms.[109] Christianity maintained a growth rate of 40 percent per decade over the course of centuries: in AD 40 there were only one thousand Christians; by AD 350 there were thirty-four million Christians, constituting over 50 percent of the population of the Roman Empire.[110] This exponential growth explains, in significant part, why Christianity became one of the major world religions. But to pagan Romans in AD 40, the idea that Christianity would become the dominant religion must have been laughable. Some modern religious groups have matched the growth rates of the early Christians. For example, in the twentieth century, the Mormon population grew at 43 percent per decade because of high fertility rates, missionary activities, and high retention.[111]

This same force will continue to shape the future. I live in an extremely secular bubble, and my naive view was that the proportion of the world which is atheist would inexorably increase. But this is not what's projected to happen. On average, atheists have few children compared to the religious, especially fundamentalists and those in poorer countries. Over time, this matters. According to the Pew Research Center, by 2050 the proportion of people with no religious affiliation (which includes atheists, agnostics, and people who do not identify with any religion but who may hold some religious or spiritual beliefs) will decrease from 16 percent to 13 percent of the world population.[112] The primary reason for this is the higher fertility rate among religious groups; conversions in and out of a religion play a surprisingly small role in total numbers.[113] If these trends continue into the future, then secular influence will slowly ebb away. This could mean that most of the world ends up following a single religion.

Similarly, many of the most powerful countries today are powerful, in part, because of historical high fertility rates. India is currently the third-largest economy in the world in part because its population grew from around 290 million people in 1900 to almost 1.4 billion people today.[114] Even though India's fertility rate has dropped to 2.2 births per woman today, it could well become the world's largest economy by the end of the century; by then, its population size is predicted to be 40 percent greater than China's.[115] For similar reasons, Nigeria looks set to become a far more important geopolitical actor by 2100 because its population is projected to grow from 200 million to 730 million, making it the third most populous country in the world.[116]

The mechanisms I've discussed so far concern competition between groups. But cultural competition also occurs between specific cultural traits, both within a culture and across cultures. The recent successes of the gay rights movement and, subsequently, the LGBTQ+ rights movements are examples of what once were minority attitudes to sexual orientation and gender identity successfully becoming much more prevalent over time. The rise of meditation and mindfulness in Western countries, and the rise of fast food in Eastern countries, are examples of specific cultural traits successfully transmitting from one culture to another.

If the world converged on a single value system, there would be much less pressure for those values to change over time. This global convergence could therefore lead to even greater persistence of values than we've seen historically. A single global value system could persist for thousands of years. And if it lasted until the development of AGI, then it could persist forever.

How Locked-In Are We Already?

I've discussed various ways in which a single value system could become globally dominant and ways in which it could entrench itself for a very long time. I've presented this as a threat we've avoided to date and will face in the future. But lock-in is not an all-or-nothing thing—there are countless distinct moral norms, each of which could be locked in or not. So we should ask about the degree to which history has already locked in certain values, or at least has made some values very unlikely to change in the future.

It's plausible to me that quite a bit of lock-in has already occurred. This starts at least with the emergence of *Homo sapiens*, which was probably analogous to the cultural lock-in that I've sketched in this chapter: a single species was able to gain power more rapidly than others and thereby entrench dominance on the planet. The members of the *Homo* genus that went extinct soon after *Homo sapiens* entered their terrain include the Neanderthals, the Denisovans, *Homo luzonensis*, *Homo erectus, Homo heidelbergensis*, and *Homo floresiensis*.[117] Now that all the other *Homo* species are extinct, there's essentially no chance that they will be resurrected and take over the world.

If evolution had gone down a different track, it's plausible that some other species, in some ways quite different from us, could have evolved cumulative cultural learning and higher intelligence. Perhaps they could have been more hierarchical, like chimpanzees, or more egalitarian, like bonobos. They could have been more aggressive, or less. They could have had more differences between the sexes, or fewer. Our biological nature leaves an awful lot open, but it still makes some sets of values more likely to thrive than others.

A second major point of lock-in, it seems to me, occurred with colonialism. *Homo sapiens* was geographically united when it evolved; then, after spreading across the world, it was separated into distinct populations. After the colonial era, the world became globally interconnected once again, so it became possible for a single ideology to have global reach. And indeed Western European powers killed off many alternative cultures, such as the Taino in the Americas, and forced their culture onto many others.[118] It resulted in the enormous spread of Christianity, of the English and Spanish languages, and of Western European culture more broadly. Since that point, because of globalization, most countries have been becoming more culturally Western over time.[119] If this process continues, there will eventually be even greater homogenisation across cultures.

One way of gauging the current diversity of cultures is to consider the range of responses countries made to the COVID-19 pandemic.[120] There was, of course, some diversity, from the ultrastrict lockdowns in China to the more moderate response in Sweden. But the range of responses was far more limited than it could have been. For example, both the Moderna and the Pfizer-BioNTech vaccines were designed by mid-January 2020 over the course of a few days.[121] Not a single country allowed human challenge trials of the many vaccines developed in 2020, where willing volunteers would be vaccinated and then deliberately infected with the coronavirus in order to very quickly test the vaccine's efficacy. Not a single country allowed the vaccine to be bought on the free market, prior to testing, by those who understood the risks, even on the condition that they report whether they were subsequently infected.[122]

I'm not going to argue here that any particular policy was better than another. But the global benefits of a diversity of responses would have been immense. If just one country had allowed human challenge trials or had allowed vaccines to be sold freely, we all would have gained the knowledge that the vaccines were effective months earlier than we did. It would still have taken significant time to ramp up production of the vaccines, but we could have brought forward the end of the pandemic by several months. In this case, homogeneity in the global response to COVID-19 was responsible for millions of deaths.

Building a Morally Exploratory World

The lock-in of some values, like Nazi or Stalinist values, would obviously have been horrific. Illustrations of some of these scenarios have been sketched in fiction. Most famous is George Orwell's *1984*, in which this bleak prospect is epitomised in the famous metaphor of "a boot stamping on a human face—forever." Even more impressive, in my view, is *Swastika Night*, written by Katharine Burdekin. It takes seriously Hitler's claim that he would create a thousand-year Reich: set seven hundred years in the future, it depicts a world which is entirely controlled by the Nazis and the Japanese Empire. In the German Empire, non-Germans have been subjugated, violence is glorified, and women are kept in pens and raped at will. To us, it reads like a piece of alternative history, but it was really a prophetic warning about ideological lock-in; the book was written in 1935, four years before World War II broke out, and published in 1937, twelve years before *1984*, at a time when Hitler still had considerable international prestige.[123]

From what I've said so far, you might conclude that we should aim to lock in the values we, today, think are right, thereby preventing dystopia via the lock-in of worse values. But that would be a mistake.[124] While the lock-in of Nazism and Stalinism would have been nightmarish, the lock-in of the values of *any* time or place would be terrible in many respects. Think, for example, of what the world would be like if Western values of just two and a half centuries ago had been locked in. The future would be shaped by values in which slavery was permissible, there was a natural hierarchy among races, women were second-class citizens, and most varieties of sexual orientation and activity were abhorrent.

Almost all generations in the past had some values that we now regard as abominable. It's easy to naively think that one has the best values; Romans would have congratulated themselves for being so civilised compared to their "barbarian" neighbours and in the same evening beaten people they had enslaved or visited the Colosseum to watch the disembowelment of a prisoner. It is extraordinarily unlikely that, of all generations across time, we are the first ones to have gotten it completely correct. The values you or I endorse are probably far from the best ones.

Moreover, there are so many ethical questions to which we *know* we haven't yet figured out the answer. Which beings have moral status: just *Homo sapiens*, or all primates, or all conscious creatures, including artificial beings that we might create in the future? How should we weigh the promotion of happiness against the alleviation of suffering? How should we handle uncertainty about the impact of our actions, especially when it comes to tiny probabilities of enormous payoffs? How should we act when we know we don't know what the right thing to do is?

And the list I've given only points to the areas of uncertainty that we know about. For thousands of years, the permissibility of slaveholding was almost unquestioned by those who dedicated their lives to ethical reflection. We should also worry about gross moral errors that we haven't yet even considered, that are invisible to us, like water to a fish.

The track record of past moral errors suggests that we are guilty of such grave errors today. We see historical attempts by the Qin, the European colonialists, and the Nazis to lock in their ideologies as terrifying, and rightly so. But if we are guilty of gross moral errors ourselves, then locking in our present values would also be a disaster.

Instead, we should try to ensure that we have made as much moral progress as possible before any point of lock-in. Political philosophers often argue over what an ideal state would look like. I think we should accept that we don't know what the ideal state would be; the primary question is how we can build a society such that, over time, our moral views improve, people act more often in accordance with them, and the world evolves to become a better, more just place.

As an ideal, we could aim for what we can call the *long reflection*: a stable state of the world in which we are safe from calamity and we can reflect on and debate the nature of the good life, working out what the most flourishing society would be. I call this the "long" reflection not because of how long this period would last but because of how long it would be *worth* spending on it. It's worth spending five minutes to decide where to spend two hours at dinner; it's worth spending months to choose a profession for the rest of one's life. But civilisation might last millions, billions, or even trillions of years. It would therefore be worth spending many centuries to ensure that we've really figured things out before we take irreversible actions like locking in values or spreading across the stars.

It seems unlikely to me that anything like the long reflection will occur. But we can see it as an ideal to try to approximate. What we want to do is build a morally exploratory world: one structured so that, over time, the norms and institutions that are morally better are more likely to win out, leading us, over time, to converge on the best possible society.[125] This would involve several things.

First, we would need to keep our options open as much as possible. This gives us a reason, though not necessarily a decisive reason, to delay events which risk value lock-in. Such potentially irreversible events might include the formation of a world government, the development of AGI, and the first serious efforts at space settlement.

It also gives us a reason to prevent smaller-scale lock-ins—for example, by supporting conservation efforts. Even if we don't know whether some species or work of art or language is valuable, there is an asymmetry between preserving it and letting it be destroyed. If we preserve it and conclude later that it's not worth holding on to, then we can always change our minds. If we let it be destroyed, we can't ever get it back.

Second, a morally exploratory world would favour *political experimentalism*—increasing cultural and intellectual diversity, if possible. We saw that we might already be on the way to a single global culture. If we are aiming to get to the best possible society, we should worry about premature convergence, like a teenager marrying the first person they date.

In *On Liberty*, John Stuart Mill argues that we should allow individual liberty and free expression because doing so creates a marketplace of ideas, where different ideas can compete and the best ideas win. We can apply the same ideas at the level of societies. The abolition of slavery came about, in part, as a result of cultural experimentation. In the eighteenth century, the United States was, comparatively speaking, a melting pot of cultural and religious diversity. This diversity enabled one community, the Quakers, to develop their own views on the morality of slavery; after they had come to see its immorality, that idea had the potential, under the right conditions, to spread.

One particularly interesting idea for promoting cultural diversity of societies is that of charter cities: autonomous communities with laws different from their surrounding countries that serve as laboratories for economic policies and governance systems. For example, in 1979 Deng Xiaoping created a special economic zone around the city of Shenzhen,[126] giving it more liberal economic policies than the rest of China. Average yearly income grew by a factor of two hundred over forty years.[127] Its success inspired broader economic reforms across China, which, over the course of the last forty years, have lifted hundreds of millions of people out of poverty.[128]

Charter cities are often promoted by those who want to see more economically liberal policies. But there is no necessary connection between these two ideas. For almost every social structure we can imagine, we could have a charter city based on that idea; there could be Marxist charter cities and environmentalist charter cities and anarchist communitarian charter cities. We could find out, empirically, which of these brings about the best society. And, in addition to creating a diversity of formal institutions, we could try to cultivate a diversity of cultures, too.

Third, we would want to structure things such that, globally, cultural evolution guides us towards morally better views and societies. I've already described a number of mechanisms by which some cultures or specific cultural traits can win out over time. Some of these mechanisms are probably not correlated with what's morally best. That one society has greater fertility than another or exhibits faster economic growth does not imply that that society is morally superior. In contrast, the most important mechanisms for improving our moral views are reason, reflection, and empathy, and the persuasion of others based on those mechanisms. If two groups engage in good-faith debate and one is convinced to change their mind via the force of reason or empathy, then, in general, that group is more likely to have gotten to an improved point of view.

Certain forms of free speech would therefore be crucial to enable better ideas to spread. Spaces for good-faith debate and careful argument and deliberation, especially, should be actively encouraged. But this is an instrumental justification of free speech, and it might not apply to all forms of speech. It seems that techniques for duping people—lying, bullshitting, and brainwashing—should be discouraged, and should be especially off limits for people in positions of power, such as those in political office. Otherwise the world could end up converging on the ideas that are most alluring rather than those that are best justified.

Fairly free migration would also be helpful. If people emigrate from one society to another, that gives us at least some evidence that that latter society is better for those who migrated there. Of the world's adults, 15 percent would like to move to another country if they had the opportunity. Demand is especially high in low-income countries, and among people who would like to move, the majority would like to move to a handful of rich liberal democracies.[129] Plausibly, this is because living in the rich liberal democracies would provide a higher quality of life.

Fairly free migration would help people to "vote with their feet," and the societies that are more attractive to live in would be rewarded with greater net immigration and grow more powerful over time. At the same time, we would want to prevent any one culture from becoming so powerful that it could conquer all other cultures through economic or military domination. Potentially, this could require international norms or laws preventing any single country from becoming too populous, just as antitrust regulations prevent any single company from dominating a market and exerting monopoly power.

This last point—that we need to structure global society so that cultural evolution guides the world towards better values and better societal structures—highlights an issue facing the design of a morally exploratory world that I'll call the *lock-in paradox*. We need to lock in some institutions and ideas in order to prevent a more thoroughgoing lock-in of values. One challenge is that these institutions and ideas will be morally controversial; for example, from many fundamentalist religious perspectives, the idea that we would encourage or even allow a diversity of worldviews might be regarded as abominable. Similarly, the idea that the path to the correct moral view is via reflection and good-faith debate, rather than studying the scripture of a holy book, is not one that everyone would accept.[130]

The lock-in paradox thus resembles the familiar paradox of tolerance—the necessity for liberal societies to defend themselves against intolerant views that would undermine their freedom, even if doing so requires curtailing the very tolerance they want to preserve.[131]

I think we must live with these paradoxes. If we wish to avoid the lock-in of bad moral views, an entirely laissez-faire approach would not be possible; over time, the forces of cultural evolution would dictate how the future goes, and the ideologies that lead to the greatest military power and that try to eliminate their competition would suppress all others.[132]

In this chapter, I've suggested that we are living through a period of plasticity, that the moral views that shape society are like molten glass that can be blown into many different shapes. But the glass is cooling, and at some point, perhaps in the not-too-distant future, it might set. Whether it sets into a sculpture that is beautiful and crystalline or mangled and misshapen is, in significant part, up to us. Or perhaps, when the glass sets, we get no shape at all; perhaps instead it cracks and shatters. Perhaps in the not-too-distant future, history ends in a more literal sense than we've discussed in this chapter: not with the victory of a single ideology, but with the permanent collapse of civilisation. It's this possibility that I'll turn to next.

PART III
SAFEGUARDING CIVILISATION

CHAPTER 5

Extinction

Spaceguard

At 09.46 GMT on the morning of 11 September, in the exceptionally beautiful summer of the year 2077, most of the inhabitants of Europe saw a dazzling fireball appear in the eastern sky. Within seconds it was brighter than the sun, and as it moved across the heavens—at first in utter silence—it left behind it a churning column of dust and smoke.

Somewhere above Austria it began to disintegrate, producing a series of concussions so violent that more than a million people had their hearing permanently damaged. They were the lucky ones.

Moving at fifty kilometres a second, a thousand tons of rock and metal impacted on the plains of northern Italy, destroying in a few flaming moments the labour of centuries. The cities of Padua and Verona were wiped from the face of the earth; and the last glories of Venice sank for ever beneath the sea as the waters of the Adriatic came—thundering landwards after the hammer-blow from space.

Six hundred thousand people died, and the total damage was more than a trillion dollars. But the loss to art, to history, to science—to the whole human race, for the rest of time—was beyond all computation. It was as if a great war had been fought and lost in a single morning; and few could draw much pleasure from the fact that, as the dust of destruction slowly settled, for months the whole world witnessed the most splendid dawns and sunsets since Krakatoa.

After the initial shock, mankind reacted with a determination and a unity that no earlier age could have shown. Such a disaster, it was realized, might not occur again for a thousand years—but it might occur tomorrow. And the next time, the consequences could be even worse.

Very well; there would be no next time.

Thus begins Arthur C. Clarke's *Rendezvous with Rama*, a science fiction novel published in 1973. In this story, the government of Earth, shaken by the asteroid strike in Italy, sets up a system called Spaceguard, an early-warning system for Earth-bound threats from space.

For years, many scientists warned of the dangers that asteroids pose to life on Earth, but for many years they weren't listened to. Even after it was first proposed, in 1980, that the dinosaurs were killed off by a huge asteroid striking the Yucatán Peninsula in Mexico,[1] there was, in the words of leading astronomer Clark R. Chapman, a "giggle factor" associated with the risk from asteroids.[2]

This all changed in 1994 when comet Shoemaker-Levy 9 thudded into the side of Jupiter with the force of three hundred billion tonnes of TNT, equivalent to 125 times the world's nuclear arsenal.[3] One of the Shoemaker-Levy fragments left a scar on Jupiter twelve thousand kilometres across, about the size of Earth.[4] David Levy noted that the comet that he codiscovered "killed off the giggle factor."[5] The impact made headlines across the world.[6] In 1998, two blockbuster films, *Deep Impact* and *Armageddon*, explored how the people of Earth might respond to a huge approaching asteroid. Scientists commended *Deep Impact* for its understanding of the impact threat and the realism of its special effects, which reflected the input of a fleet of technical advisers that included Gene Shoemaker, whom the comet Shoemaker-Levy was named after.[7] (*Armageddon*, in contrast, was described by Clark Chapman as "scientifically and technologically preposterous in almost every respect."[8])

Due to increasing interest from the public and advocacy from scientists, in 1998 Congress tasked NASA with finding 90 percent of all near-Earth asteroids and comets larger than one kilometre within a decade.[9] The effort would, with due acknowledgement to Arthur C. Clarke, be called Spaceguard.[10]

Spaceguard has been a huge success. We have now tracked 93 percent of asteroids larger than one kilometre and found more than 98 percent of the extinction-threatening asteroids, which measure at least ten kilometres across.[11] Prior to Spaceguard, the estimated risk that Earth would be hit by an extinction-level asteroid was around one in two hundred million per year.[12] We know now that the risk is less than one in fifteen billion—one hundred times lower.[13]

The last two chapters discussed ways that we can make the future better, for however long civilisation lasts. This chapter and the next two will look at ways we can ensure that we have a future at all, beginning with how to avoid the near-term extinction of our species.

Spaceguard showed that we have what it takes to manage risks to the extinction of humanity, if we put our mind to it. Though we discovered that there was no imminent threat from asteroids, the tracking meant that if we *had* discovered an asteroid on course to collide with Earth, we could have devoted enormous resources to deflecting it and to building food stockpiles in case we failed. A few hundreds of millions of dollars was enough to appropriately manage this risk.[14] But in the coming decades, we will have to deal with much greater risks. If we do not rise to the challenge, there is a decent chance that humanity could come to a premature end and our future could be destroyed.

Engineered Pathogens

Most of this book was written during the COVID-19 pandemic. At the time of writing, COVID-19 is estimated to have caused seventeen million excess deaths worldwide—one in every five hundred people.[15] The death toll is sure to increase in the future. The economic cost will amount to more than $10 trillion.[16] And billions of people have lived under lockdown for months on end, unable to see their family and friends in person, even when dying in hospital.

But, despite the toll of COVID-19, in some respects we've gotten off easily. We know that viruses (like Ebola) can be deadlier than the new coronavirus, and some (like the measles) can be more transmissible. If the new coronavirus had been ten times as deadly, then the death toll could have amounted to hundreds of millions or more.

Looking to the future, the threat posed by pandemics may be much greater still. This greater threat comes not from naturally arising pathogens but from diseases that we ourselves will design, using the tools of biotechnology.

Biotechnology is an area of research that tries to build new biological entities or alter those already found in nature. Progress in this field has been extremely rapid. We typically think Moore's law—halving the cost of computing power every few years—is the prime example of quick progress, but many technologies in synthetic biology actually improved faster than that.[17] For example, the first time we sequenced the human genome, it cost hundreds of millions of dollars to do so. Just twenty years later, sequencing a full human genome costs around $1,000.[18] A similar story is true for the cost to synthesise single-strand DNA, as well as the cost of gene editing.

This rapid technological progress promises great benefits in medicine and in the treatment of rare genetic diseases, but it also brings unprecedented risks, in particular because it gives us the power to design and create new pathogens.

Engineered pathogens could be much more destructive than natural pathogens because they can be modified to have dangerous new properties. Could someone design a pathogen with maximum destructive power—something with the lethality of Ebola and the contagiousness of measles? Thankfully, with current technology this would be at least very difficult. But given the rate of progress in this area, it's only a matter of time.

Not only is biotechnology rapidly improving; it is becoming increasingly democratised. The genetic recipe for smallpox is already freely available online.[19] In a sense, we were "lucky" with nuclear weapons insofar as fissile material is incredibly hard to manufacture. The capability to do so is therefore limited to governments, and it is comparatively easy for outside observers to tell whether a country has a nuclear weapons programme.[20] This is not so for engineered pathogens: in principle, with continued technological progress, viruses could be designed and produced with at-home kits. In the future, cost and skill barriers are likely to decline. Moreover, in the past we only had to deal with one pandemic at a time, and usually some people had natural immunity; in contrast, if it's possible to engineer one type of new highly destructive pathogen, then it's not that much harder to manufacture hundreds more, nor is it difficult to distribute them in thousands of locations around the world at once.

Since the techniques of biological engineering are becoming ever more powerful and ever more democratised, one would hope that there would be a commensurate improvement in caution and safety around this research. We would expect laboratories doing this research to have extremely high safety standards and the research to be very strictly regulated, with severe punishment for any lapses in safety. But in fact, the level of biosafety around the world is truly shocking. For example, I remember as a teenager seeing images on the news of giant pyres burning thousands of cow carcasses. This was reporting of the 2001 UK outbreak of foot-and-mouth disease, an infection affecting hooved animals that causes a high fever and painful blisters in the mouth and feet and sometimes leads to lameness and death. The outbreak originated with pigs that were fed garbage containing the remains of illegally imported meat contaminated with the disease, and it spread to over two thousand farms across the UK.[21] Before it was finally contained, the outbreak led to the culling of millions of sheep and cattle and cost a total of £8 billion.[22] After it was contained, enormous effort went into making sure it did not happen again: government reports were written; laws were changed.[23]

But just six years later, there was another foot-and-mouth outbreak. Unlike the 2001 outbreak, the 2007 outbreak started with a leak from a lab that was developing vaccines to protect livestock against foot-and-mouth disease.[24] Some of the pipes carrying waste from the lab to the facility's waste treatment were old and leaky, and waste contaminated with the disease leaked out into the soil and eventually reached a nearby farm.[25] The poor maintenance of those pipes constituted a clear violation of the lab's licence to work with an infectious pathogen.[26] While this outbreak was caught and contained within weeks, it never should have happened in the first place.[27]

So after this disaster, surely the utmost precautions were taken to prevent the risks of a foot-and-mouth outbreak happening again, right? Sadly, no. Soon after the containment of that 2007 outbreak, there was a third outbreak, just a few weeks later, from the very same lab. The lab had failed to comply with the government's conditions for resumption of their vaccine production and once again leaked foot-and-mouth into the environment.[28]

These are not isolated events; in fact, uncontrolled pathogen escapes are almost commonplace. In one of the deadliest confirmed lab leaks on record, over one hundred people died after being exposed to anthrax 836, the most powerful strain of anthrax in the Soviet bioweapons programme, in April

1979.[29] A technician in a covert anthrax-drying plant in the city of Sverdlovsk removed a clogged filter without replacing it. He scribbled a note for his supervisor but forgot to record it in the logbook; his supervisor didn't find the note and started up the plant, and anthrax escaped through the filterless vent and was carried to nearby buildings by the wind.[30] In another instance, in 1971, a woman on an environmental research ship in the Aral Sea was exposed to a strain of smallpox that was probably used in a nearby bioweapon field test.[31] The strain had been designed to be highly virulent and possibly vaccine-resistant, and it was aerosolised so that it could travel across large distances.[32] While she was still asymptomatic, she returned to her home city of Aralsk, where nine others subsequently became infected, including a woman and two children who then died.[33] Soviet officials locked Aralsk down, incinerated several properties, and vaccinated the entire population of fifty thousand people, preventing a larger outbreak of one of the deadliest viruses in the world, but perhaps only narrowly.[34]

Similarly, smallpox leaked from virology labs not once but three times in the UK during the 1960s and 1970s. A mild strain infected a medical photographer working above an unsafe virology lab at the University of Birmingham in 1966, leading to seventy-two confirmed cases.[35] In 1973, a lab technician at the London School of Hygiene and Tropical Medicine became infected with smallpox and was then placed in an open ward where he infected two people visiting a patient in an adjacent bed; the two visitors' infections were fatal.[36] In fact, the last person to ever die from smallpox, Janet Parker, who died in 1978, was a medical photographer working above the very same Birmingham lab that had caused the 1966 outbreak.[37] And between 1979 and 2009, there were 444 infections in labs permitted to work with especially dangerous pathogens.[38]

The accidents were caused by a mix of human error and equipment failures and involved diseases like Ebola, anthrax, Rift Valley fever, and encephalitis.[39]

Even if it becomes possible to build pathogens that are far more destructive than foot-and-mouth or COVID-19, surely no one would *want* to do so? After all, bioweapons seem useless for warfare because it's extremely difficult to target who is infected. If you create a virus to decimate the opposing side, it's likely that the pandemic will invade your home country too.

One can think up counterarguments. Perhaps, for example, the country deploying the bioweapons would first vaccinate its population against them; perhaps, as a deterrent, the country would create an automated system guaranteed to release such pathogens in the event of a nuclear attack.[40] But the stronger counterargument is that, as a matter of fact, major bioweapons programmes *have* been run.

In the past, the United States, Japan, and the Soviet Union all had major bioweapons programmes.[41] The Soviets' was most extensive by far, lasting sixty-four years and employing as many as sixty thousand personnel at its height.[42] They built entire cities, not found on any map and not accessible to foreigners, where they did all of their bioweapons research.[43] While most other countries' bioweapons programmes were limited in both their scope and success, the Soviet programme managed to develop a wide range of bioweapons that could assassinate individuals, kill crops, and even incapacitate people across large areas, though these weapons were not operationally useful.[44] The programme was highly secretive. While the USSR claimed to have shut down its bioweapons programme in 1972 when it signed the Biological Weapons Convention, it continued running it until the collapse of the Soviet Union; in fact, it is unclear whether Russia has ever completely dismantled the Soviet programme.[45] The programme was not known to the United States until the Russians voluntarily disclosed information about it in 1991, though it had been suspected earlier because of defector accounts and the anthrax outbreak at Sverdlovsk.[46]

Even if such weapons are never used in warfare in the future, they could still leak from the labs where they are developed. The list of lab escapes I discussed before only includes those that have been confirmed. The true number is probably much higher. Data on infections that have happened in US labs that work with relatively dangerous pathogens indicate that for every year that 250 full-time employees are working in these labs, there has been one accidental infection.[47] If we assume the Soviet bioweapons programme saw accidental infections at the same rate as US labs, then we should expect that there were thousands of lab-escape infections from the Soviet programme.[48] And that assumes the Cold War–era Soviet bioweapons programme was as cautious as the post–Cold War US biomedical community. Instead, it was probably much riskier.[49] Given the lengths the Soviet Union went to to keep their bioweapons programme secret, it seems possible that they kept thousands of accidental lab infections secret as well. After all, they managed to conceal the outbreaks from their bioweapons programmes in Sverdlovsk and Aralsk.[50] Supporting this theory are cases where there is at least some evidence that disease outbreaks thought to have come about naturally may have actually been the result of human error. For example, there is now some evidence, based on genetic analysis, that the 1977 Russian flu pandemic, which according to one estimate killed seven hundred thousand people, may have either leaked from a lab or resulted from a poorly implemented vaccine trial.[51]

I think it is difficult to rule out the possibility that synthetic biology could threaten human extinction. One could try to approach this problem by anticipating specific ways novel technology could be misused. However, in doing so one would need to carefully balance the risk mitigation benefits of improved foresight against the risks of lab accidents and inspiring bad actors. There is some precedent for the latter. For example, starting in 1927, Major Shiro Ishii spent years lobbying the reluctant Japanese Ministry of War to pursue a bioweapons programme. He learned about the power of bioweapons after reading about them in a Japanese physician's report on the 1925 Geneva Disarmament Conference—a convention whose key purpose was to garner support for a ban on chemical and biological weapons. He successfully convinced the Japanese military to pursue a bioweapons programme, arguing that biological warfare must be worth pursuing, "otherwise, it would not have been outlawed by the League of Nations."[52] Now infamous for its extensive experimentation on human subjects, the Japanese bioweapons programme existed for eleven years and grew to employ a few thousand personnel.

Similarly, the man who conceived of al-Qaeda's bioweapons programme, Ayman al-Zawahiri, wrote that he had only become aware of their destructive power after "the enemy drew our attention to them by repeatedly expressing concern that they can be produced simply."[53] After they invaded Afghanistan, the United States found books and journal papers relevant to building bioweapons and plans for a bioweapons lab in an al-Qaeda training camp near Kandahar. The documents also showed that an al-Qaeda operative with doctoral training in microbiology had tried to acquire bioweapons and vaccines for workers at the planned laboratory.[54] By sounding the alarm bell, we risk making it more likely that such a catastrophe could occur.

Yet for risk mitigation, it is important to understand which dangers to our future loom largest. Many extinction risk specialists consider engineered pandemics the second most likely cause of our demise this century, just behind artificial intelligence. At the time of writing, the community forecasting platform Metaculus puts the probability of an engineered pandemic killing at least 95 percent of people by 2100 at 0.6 percent.[55] Experts I know typically put the probability of an extinction-level engineered pandemic this century at around 1 percent; in his book *The Precipice*, my colleague Toby Ord puts the probability at 3 percent.[56]

Even if you dispute the precise numbers, I think that in no way can we rule out such a possibility. And even if the probability is low, it is still high enough that preventing such a catastrophe should be a key priority of our time. Imagine you were stepping aboard a plane and you were told that it had "only" a one-in-a-thousand chance of crashing and killing everyone on board.[57] Would you feel reassured?

Only once a huge comet collided with a nearby planet, creating a fireball that reached more than thirty thousand degrees Celsius,[58] did governments and the wider public turn their attention towards the risk from asteroids and comets. It is tragic that it might take something as disastrous as COVID-19 to convince the world to pay more attention to pandemics. And the COVID-19 pandemic is tame in comparison with the horrors that novel engineered pathogens might bring. The world eventually got its act together on asteroids and comets. It is time we did the same for engineered pathogens.[59]

People in the longtermist community were warning about pandemics for many years prior to COVID-19. One of the main longtermist funders, Open Philanthropy, was one of the few pre-COVID funders of pandemic preparedness in the world. It made its first grant in the area in 2015 and has since given out more than $100 million in the area. The group 80,000 Hours has recommended careers in pandemic preparedness since 2016. In 2017, I had dinner with Nicola Sturgeon, the first minister of Scotland, and was given the opportunity to pitch her on one policy. I chose pandemic preparedness, focusing on worst-case pandemics. Everyone laughed, and the host of the dinner, Sir Tom Hunter, joked that I was "freaking everyone oot."

Great-Power War

The greatest driver of engineered pathogens so far was undoubtedly the Cold War. In the hunt for military superiority, the Soviets pursued a bioweapons programme that achieved nothing except the deaths of dozens of Russians and the exposure of millions more to the risk of a horrific death. Simply put, when people are at war or fear war, they do stupid things.

Wars are tragic no matter where and when they happen, but especially concerning from a longtermist perspective are those that pit the most powerful countries of their time—the "great powers"—against each other. This is simply because of the sheer scale of destructiveness required to cause human extinction or other irrecoverable harms to future generations: an all-out war between the world's largest and most technologically advanced militaries is more likely to exceed that grim threshold than more limited conflicts.

Longtermists may thus be tempted to rejoice in the observation that soldiers from the great powers haven't met in battle since the end of World War II. This "Long Peace" might suggest that great-power wars are a relic of the past, or at least much less likely today.[60]

Unfortunately, I don't think we can take the Long Peace for granted. As I revise this chapter for publication, Russia's invasion of Ukraine reminds us that war can all too quickly return to regions that have enjoyed peace for decades, and that initially more limited disputes can push the world's largest nuclear powers dangerously close to the brink of a direct confrontation. And there are several reasons to think that the risk of great-power war in the next hundred years remains unacceptably high.

First, it seems plausible that maintaining the Long Peace has involved a healthy dose of luck in addition to structural factors like economic growth and international cooperation. We know that the United States and the Soviet Union came close to war during the Cuban Missile Crisis, for example. But this was hardly the Cold War's only moment of danger. Tensions were also high during the Berlin crises, the Suez Crisis, the 1973 Arab-Israeli war, multiple crises in the Taiwan Strait, and proxy wars in Korea and Vietnam, as well as on several occasions when early-warning systems failed and sent false alarms of incoming nuclear attacks.[61] World War II has been characterized in part as hugely unlucky, due to Hitler's unlikely rise.[62] But the peace that followed has also been partly the result of chance.[63]

Second, changes in the distribution of global economic and military power may increase the risk of conflict. China is on track to surpass the United States on a number of dimensions. Indeed, after adjusting for purchasing power, China's economy is already larger than the United States'.[64] Power transition periods, when one superpower nation surpasses another, appear to be especially unstable times as rival powers compete for influence over the international system.[65] While war is far from inevitable at such times, and many past power transitions have been peaceful, several scholarly analyses have found that the proportion of transitions that do turn violent is worryingly high.[66]

Many different factors contribute to the decision to go to war, but disagreement over relative status and the distribution of political, economic, and military power within the international system can play important roles.[67] Alliance commitments can draw distant countries into regional disputes. Powerful countries and countries that have long-standing rivalries are more likely to fight than other countries.[68]

The United States and China are poised to be the most powerful countries by far in the coming decade, but there are significant risks of war between other great powers too. Russia maintains an enormous arsenal of nuclear warheads,[69] and the US-Russia relationship has deteriorated. India is projected to be the most populous country in the world by 2030 and could overtake China as the world's largest economy this century.[70] There are also significant military tensions between India and China. While writing this book, I read the news about the Galwan Valley clash on June 15, 2020—a violent skirmish between Indian and Chinese soldiers in territory high in the Himalayan mountains that is claimed by both countries. The two countries had made agreements not to use firearms along the disputed border, so instead, they attacked each other with stones, clubs, and batons wrapped in barbed wire. More than twenty people died.[71] One report suggested that "ties between both countries [had] reached their lowest point since the 1962 [Sino-Indian] war."[72]

To be clear, war between great powers this century is not inevitable. For one, power transitions do not inexorably end in conflict. In the twentieth century the United States surpassed Great Britain, and the Soviet Union became a major force in Eurasia, without these countries coming into direct conflict with each other. And the US-China relationship, at least, lacks some of the characteristics of the most dangerous kind of international rivalry. In particular, the countries do not share a border or claim any of the same territory, two powerful factors that push countries towards war.[73] Their economies are also entwined, as each is currently among the other's largest trading partners, which some researchers think makes war more costly and, hence, less likely.[74] Finally, if the last seventy years of peace have been the result of systemic, enduring changes to the way countries relate to each other, then peace may continue. Perhaps a nuclear war would be so destructive for everyone involved that it's not worth taking any actions that risk causing one.[75] Some scholars also think that the prospect of deploying nuclear weapons seems so wrong that their use has become taboo.[76]

Following Russia's invasion of Ukraine, the community forecasting platform Metaculus more than doubled its predicted chance of a third world war by 2050, to 23 percent (defining a world war as one involving countries representing either 30 percent of GDP or 50 percent of world population and killing at least ten million people).[77] If that annual risk stayed the same for the following fifty years, this would mean another world war before the end of the century is more likely than not. What makes this especially troubling is that growing military spending and new technologies are increasing humanity's capacity to wage war. If the great powers came to blows in the future, they could deploy weapons far more destructive and lethal than those used in World War II. The potential for devastation is enormous.[78]

Just as smoking increases the risk of practically all forms of cancer, great-power war also increases the risk of a host of other risks to civilisation. First, it diverts spending away from things that improve the safety and quality of life, and second, it destroys our ability to cooperate. The Cold War led the Soviet Union to the insanity of a secret bioweapons programme; a new conflict between the major powers would increase the temptation to develop new biological weapons of mass destruction. Even if it didn't lead to direct, violent conflict, a new Cold War could also increase the risk of an AI arms race and so increase the risk of bad-value lock-in or misaligned AI takeover. It would increase the risk that nuclear weapons are used, and it would undermine our ability to cooperate internationally to deal with climate change. In my view, reducing the likelihood and severity of the next world war is one of the most important ways we can safeguard civilisation this century.

Would a Technologically Capable Species Re-evolve?

For human extinction to be of great longterm importance, it needs to be highly persistent, significant, and contingent. Its persistence might seem obvious: if we go extinct, we can't come back from that. But there's a counterargument one could make. Even if the end of *Homo sapiens* is highly persistent, perhaps the end of morally valuable civilisation is not. That is, perhaps if *Homo sapiens* went extinct, some other technologically capable species would evolve and take our place.

The last common ancestor of humans and chimps was alive only twelve million years ago, and it took only around two hundred million years for humans to evolve from the first mammals.[79] And there are still at least hundreds of millions of years remaining until the sun's increasing brightness renders the earth uninhabitable to human-size animals. Given this, if *Homo sapiens* went extinct and chimps survived, shouldn't we expect a technologically capable species to evolve from chimps, like *Planet of the Apes,* in eight million years or less? Similarly, even if all primates went extinct, as long as some mammals survived shouldn't we expect a technologically capable species to evolve within around two hundred million years? This is a long time, but it's still easily short enough for such evolution to occur before the earth is no longer habitable.

This argument is too quick. We don't know how unlikely the major evolutionary transitions were, and there is reason to believe that some of them—including, potentially, the evolution of a technologically capable species—were very unlikely indeed.

There are two reasons to think this. The first is based on the Fermi paradox: the paradox that, even though there are at least hundreds of millions of rocky habitable-zone planets in the galaxy, and even though our galaxy is 13.5 billion years old,[80] giving ample time for an interstellar civilisation to spread widely across it, we see no evidence of alien life. If the galaxy is so vast and so old, why is it not teeming with aliens?

One answer is that something about our evolutionary history was exceptionally unlikely to occur.[81] Perhaps planets that are conducive for life are in fact extremely rare (perhaps needing to be in a safe zone in the galaxy, with plate tectonics, a large moon, and the right chemical composition), or certain steps on the path from the formation of the earth 4.5 billion years ago to the evolution of *Homo sapiens* were extraordinarily unlikely.[82] Potentially extremely improbable steps include the creation of the first replicators from inorganic matter, the evolution of simple cells into complex cells with a nucleus and mitochondria, the evolution of sexual reproduction, and possibly even the evolution of a species, like *Homo sapiens*, that is distinct from other primates by virtue of being unusually intelligent, hypercooperative, culturally evolving, and capable of speech and language.[83] Recent research by my colleagues at the Future of Humanity Institute suggests that once we properly account for our uncertainty about just how unlikely these evolutionary transitions might be, it actually becomes not all that surprising that the universe is empty, even though it is so vast.[84]

The second reason to think that one or more of the evolutionary transitions in our past were very unlikely is how long it took for *Homo sapiens* to evolve. Consider this: Suppose that, for an Earth-like planet, it should take, on average, a trillion trillion years from the planet's cooling to the evolution of a species capable of building civilisation. If this were true, what would we expect to see in our past? Well, we would expect it to look almost exactly the way our actual past does: we would have evolved fairly close to the end of the habitable lifetime of the earth. Because there are only around five billion years from the cooling of the earth to the end of the period over which it could sustain life, there's no way in which we could have evolved except by being extraordinarily lucky.[85] Because we would see the same timescales of evolutionary history whether the transition from an Earth-like planet to a technologically capable species ought to have taken five billion years or a trillion trillion years, we just can't infer how likely or unlikely that transition was.

We don't currently know how many extremely unlikely evolutionary transitions there were in our past. Some research suggests that we should expect there to have been between three and nine "hard steps" on the path to the evolution of advanced life.[86] But there has only been very limited investigation of this question, and the true number could well be higher or lower.[87] And we don't know how unlikely it was that biological evolution would produce a species that was capable of building civilisation, even after mammals or primates had evolved. For all we currently know, the evolutionary step from mammals to a species capable of building civilisation could have been astronomically unlikely to occur.

We therefore cannot be confident that, were human civilisation to end, some other technologically capable species would eventually take our place. And even if you think that there is a 90 percent chance that this would happen, that would only reduce the risk that a major catastrophe would bring about the permanent end of civilisation by a factor of ten: the risk would still be more than great enough that reducing it should be a pressing moral priority.

Moreover, if some step in our evolutionary history was extremely improbable, there might be no other highly intelligent life elsewhere in the affectable universe, and there might never be. If this is true, then our actions are of cosmic significance.

With great rarity comes great responsibility. For thirteen billion years, the known universe was devoid of consciousness; there was no entity such that, to borrow a phrase from Thomas Nagel, it *was like something to be them*. Around five hundred million years ago, that changed, and the first conscious creatures evolved: the spark of a new flame. But those creatures were not conscious of being conscious; they did not know their place in the universe, and they could not begin to understand it. And then, merely a few thousand years ago, over a little more than one ten millionth of the life span of the universe so far, we developed writing and mathematics, and we began to inquire about the nature of reality.

Now and in the coming centuries, we face threats that could kill us all. And if we mess this up, we mess it up forever. The universe's self-understanding might be permanently lost and, within just a few hundred million years more, the brief and slender flame of consciousness that flickered for a while would be extinguished forever. The universe might return eternally to the state it occupied for much of its first thirteen billion years: cold, empty, dead.

Extinction is not the only way in which civilisation might come to an end. Perhaps instead some disaster falls short of killing everyone but causes civilisation to collapse and we never recover. I'll turn to that possibility next.

CHAPTER 6
Collapse

The Fall of Empires

In AD 100, there were two major empires in the world, about equal in territorial extent and population; between them they encompassed more than one half of the world's population.[1] We discussed one of them, the Han dynasty, in Chapter 4: that was the empire that locked in Confucianism as the primary Chinese ideology for over two thousand years. The other was the Roman Empire, which had a very different fate.

If you had been the Roman emperor in AD 100, you would have regarded yourself as ruling the pinnacle of technological, legal, and economic advancement. You would have had much to support your view. Your empire enjoyed the benefits of central heating and double glazing, which insulated your public baths.[2] You used concrete which was in some ways more durable than the concrete we use today.[3] You built mighty structures, such as the Colosseum, which could seat more than 50,000 people, and the Circus Maximus, a chariot-racing stadium that could seat 150,000.[4]

You controlled an area larger than today's European Union,[5] despite having no modern means of travel such as railways or airplanes, or modern communication technology. Your economy was complex and sophisticated, with a high degree of division of labour, a banking system, and international trade across continents; traders roamed your empire selling goods and spreading knowledge.[6] There was even a historically unusual middle class of around 10 percent of the population and upward social mobility, as evidenced by satires about the follies of the "new rich."[7] Even peasants

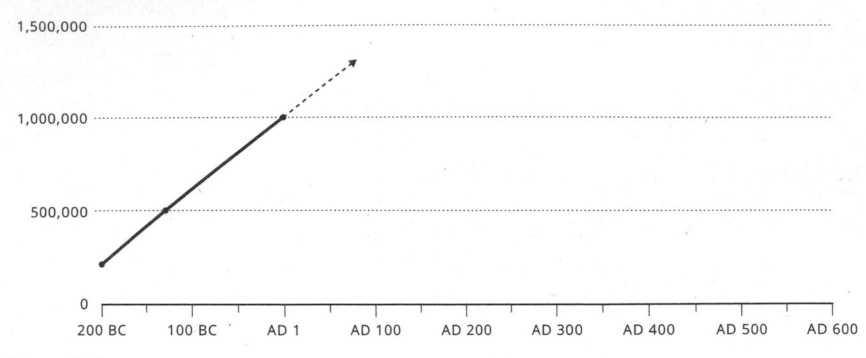

Figure 6.1.

under your rule had access to useful goods like high-quality pottery and tiled roofs.[8]

The growth of the Roman Empire's prosperity is reflected in the population growth of Rome, the first city ever to reach one million residents (see Figure 6.1).[9]

In the Roman Republic, the price of Rome's growth was the blood of its citizens and neighbours. Between 410 BC and 101 BC, Rome was at war more than 90 percent of the time.[10] After the formation of the Roman Empire in 27 BC, though, Rome experienced two centuries of growth in both population and living standards. Rome was strong and stable. At the time, it would have seemed like the city's flourishing, driven by advances in technology and governance, would continue long into the future.

This is not what happened. To illustrate this, let's look again at the graph of Rome's population but extend the timeline (see Figure 6.2).

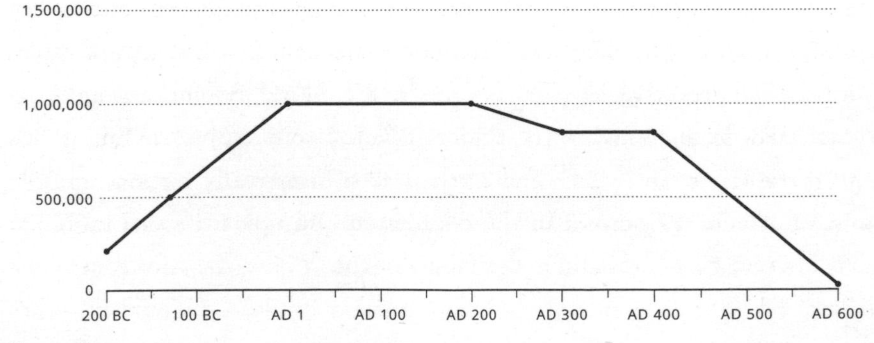

Figure 6.2.

In the fifth century, the city of Rome was sacked twice by marauding Germanic tribes: in AD 410 by the Visigoths and in AD 455 by the Vandals.

On learning of the AD 410 sack of Rome, Saint Jerome commented: "The brightest light of the whole world is extinguished; indeed the head has been cut from the Roman empire. To put it more truthfully, the whole world has died with one City. Who would have believed that Rome, which was built up from victories over the whole world, would fall; so that it would be both the mother and the tomb to all peoples."[11]

Although Rome was no longer the seat of imperial power in the Western Roman Empire at that time, the decline of the city of Rome in the fifth century vividly symbolised how weak the Western Roman Empire had become.[12] A few decades later, the whole Western Roman Empire collapsed. Rome's population dwindled to only thirty thousand people, stayed at a similar level for centuries, and only surpassed its peak population again 1,400 years later, in the 1930s.[13] In fact, it wasn't until the early nineteenth century that *any* European city surpassed the population of Rome at its ancient peak.[14]

Why, then, did the Western Roman Empire fall? A review from 1984 found that historians had suggested no fewer than 210 distinct causes for the fall of the Western Roman Empire.[15] Many modern historians agree on the basic narrative of Roman decline: flawed institutions; domestic power struggles over political position and surplus extraction; corruption and economic weaknesses; pressure from external invaders; and increasing detrimental impact of plagues and climate change.[16]

Because of the difficulty of managing a giant empire with premodern technology and communication, it is not surprising that the Roman Empire eventually crumbled, and it is more pertinent to ask why it survived for so long.[17] Indeed, the average life span of a civilisation is only around 340 years.[18] For local civilisations, collapse is the rule, not the exception.

In the last chapter, I discussed the risk of human extinction, which is one way that civilisation could come to an end. But disasters that kill everyone are very extreme; civilisational collapse and global catastrophes that fall short of killing everyone are arguably much more likely. Could the world today suffer the same fate as the Roman Empire?

I'll use the term "civilisational collapse" to refer to an event in which society loses the ability to create most industrial and postindustrial technology. If there's a good chance that such a collapse would be permanent, then the risk of civilisational collapse could be of even greater longterm importance than the risk of extinction. So let's ask: How likely is it that some nonextinction catastrophe could cause civilisation to collapse, and if it did, how likely would recovery be?

The Historical Resilience of Global Civilisation

The historical evidence suggests that human civilisation has been surprisingly resilient after catastrophe. The first thing to bear in mind is just how different a global and permanent collapse of civilisation would be from historical civilisational collapses. The fall of the Western Roman Empire is a particularly dramatic historical example of civilisational collapse. But even though Europe's mightiest empire fell, Europe was not completely depopulated. Rather, Roman rule was supplanted by the Visigoths, Vandals, Ostrogoths, Franks, Britons, and Saxons.

Still, technological sophistication and living standards did decline precipitously after the fall of Rome. Britain was an extreme case: in the fifth century, the use of writing vanished and all of the Romans' building crafts disappeared.[19] Stone, brick, and tiled buildings gave way to wood and thatch.[20] But this technological and cultural decline was not permanent. It was out of the ashes of the Western Roman Empire, centuries later, that the Renaissance, the Scientific Revolution, the Industrial Revolution, and the Enlightenment were born. Indeed, in the accounts of several leading economic historians, the comparative political fragmentation of Europe after the fall of Rome partly explains why the Scientific and Industrial Revolutions occurred there rather than in China.[21]

Moreover, all historical civilisational collapses to date have been local. When the Western Roman Empire collapsed, some of the other major civilisations of the time—such as the Northern and Southern dynasties in China, the Aksumite Empire in Ethiopia, the Three Kingdoms of Korea, Teotihuacan in Mexico, the Maya civilisation in Central America, the Sasanian dynasty in modern-day Iran, and the Gupta Empire in India[22]—continued much as before, and many of them knew nothing of the Roman Empire in the first place. Despite losing its western partner, the Eastern Roman Empire, or Byzantium, survived for another thousand years.

Indeed, even huge crises have failed to knock global civilisation off course. Over the last sixty years, the period for which we have the best data, world GDP has only shrunk in a single year a handful of times, and it has always completely rebounded within a couple of years.[23] It is not even clear whether the population declined during the Spanish flu pandemic of 1918, in which seventeen million to one hundred million people died.[24] Even though World War II was the deadliest war in history by the number of casualties, it did not cause the global population to decline.[25] The last time global population even came close to declining over a period of decades was during what some historians refer to as the "General Crisis."[26] This was a period over the seventeenth century when almost everything was going wrong: major wars in Europe, China, and India, including the Thirty Years' War and the collapse of the Ming dynasty; the widespread deaths of Indigenous Americans from European colonialism; the rise of the transatlantic slave trade; and what's called the "Little Ice Age," where temperatures in Europe cooled, leading to widespread famine.[27] The global population loss may have been large: in the first half of the seventeenth century, according to some estimates, the Chinese population plummeted by around 40 percent, while Germany and parts of France lost 20 percent to 45 percent of their populations.[28] Yet despite these crises, by AD 1700 the world population was larger than before the General Crisis.

A vivid illustration of historical societal resilience comes from the Black Death, a pandemic of the bubonic plague in the fourteenth century that spread across the Middle East and Europe. The Black Death was mainly spread by infected fleas transported across the world by rats on trade ships fleeing the Mongol invasion of Crimea. It may have been the deadliest natural catastrophe in history when measured as a percentage of world population lost. Somewhere between one-quarter and one-half of all Europeans died, and the Middle East was also terribly affected.[29] All in all, around one-tenth of the global population lost their lives.[30] Those who died did so in utter misery.

If any natural event would have brought about the collapse of civilisation, we would have expected this to be it. But, despite the enormous loss of human lives and intense suffering that the Black Death caused, it did little to negatively impact longer-term European economic and technological development. European population size returned to its prepandemic levels two centuries later; European colonial expansion continued and the Industrial Revolution occurred just four centuries later.[31]

Other examples of remarkable societal resilience are more recent. We can consider, for example, the atomic bombing of the Japanese city of Hiroshima in 1945. The bomb the United States dropped was 1,500 times more powerful than any previously used.[32] The fireball at the hypocenter of the blast reached several thousand degrees Celsius within one-ten thousandth of a second before igniting all flammable material within one and a half miles.[33] Ninety percent of the city's buildings were at least partially incinerated or reduced to rubble.[34] Initial estimates suggested that 70,000 died because of the bombing before the end of 1945, while more recent estimates put the figure at 140,000.[35] The heat from the blast was so ferocious that steps, pavements, and walls were brightened, and the people incinerated in the blast left darkened shadows. One person, thought to be a woman named Mitsuno Ochi, left a shadow on the steps of the Bank of Japan, now preserved at the Hiroshima Peace Memorial Museum in an exhibit known as the Human Shadow of Death.[36]

Before learning about Hiroshima's subsequent history, I would have thought that, even today, it would be a nuclear wasteland, consisting of little more than smoking ruins—Mitsuno Ochi's shadow on a citywide scale. But nothing could be further from the truth.[37] Despite the enormous loss of life and destruction of infrastructure, power was restored to some areas within a day, to 30 percent of homes within two weeks, and to all homes not destroyed by the blast within four months.[38] There was a limited rail service running the day after the attack, there was a streetcar service running within three days, water pumps were working again within four days, and telecommunications were restored in some areas within a month.[39] The Bank of Japan, just 380 metres from the hypocenter of the blast, reopened within just two days.[40] The population of Hiroshima returned to its predestruction level within a decade.[41] Today, it is a thriving modern city of 1.2 million people.[42]

The remarkable recovery from such unfathomable destruction is a testament to the resilience of the people of Hiroshima and the surrounding towns. But Hiroshima wasn't unique. While reconstruction was slower in Nagasaki after it was bombed, the story is fundamentally similar: Nagasaki surpassed its former population in under a decade and is now a prosperous city. And a broader study on the bombing of Japanese cities during World War II suggests that this rally was widespread. Dozens of Japanese cities had at least half of their buildings burned to the ground.[43] But these cities soon returned to their previous size, economic output, and even share of particular industries.[44]

A similar study of Vietnamese cities after the Vietnam War reached much the same conclusion. The Vietnam War involved the most intense aerial bombing in history: the US Air Force dropped on Vietnam three times the weight of bombs it used in World War II. But, remarkably, the authors of the study found no impacts of this bombing on local poverty rates, consumption levels, infrastructure, literacy rates, or population density twenty-five years after the end of the war.[45]

Sometimes people claim that, because the modern world is so complex and interreliant, it is therefore fragile, and if one strut is lost, the entire structure will fall in a domino effect. But this idea neglects people's astonishing grit, adaptability, and ingenuity in the face of adversity. This adaptability can be seen even when a disaster-struck area is cut off from the rest of the world and cannot receive assistance from elsewhere. For example, when Serbian armed forces laid siege to the city of Goražde, Bosnia, between 1992 and 1995, the city lost much of its physical infrastructure and was cut off from the national power grid. But residents of Goražde jury-rigged hydroelectric generators using scavenged alternators to meet basic power needs.[46] In an even more extreme case, after the fall of the Soviet Union, which had been the sole supplier of Cuba's agricultural equipment and supplies, Cuba lost all access to fossil fuels, fertilizers, pesticides, and agricultural machinery and depleted its stores within a few years. In response, Cuba implemented an emergency programme to breed four hundred thousand oxen to replace its industrial machinery, allowing it to avoid widespread famine.[47]

Would We Recover from Extreme Catastrophes?

Perhaps, though, the historical track record is a misleading guide to our resilience to future catastrophes. After all, we have no historical examples of global catastrophes killing more than 20 percent of the world population. But now, with nuclear weapons, we have the capacity to kill a much greater fraction of the population; advanced bioweapons will make this capacity even greater. If there were a catastrophe of unprecedented severity, would society collapse? And if it did collapse, would it ever recover?

I'll look at these questions by exploring the potential impact of an all-out nuclear war, though my analysis also applies to other catastrophes, including those involving biological weapons.

The bombings of Hiroshima and Nagasaki saw the use of weapons that were more than 1,500 times more powerful than the most powerful explosives of the time. But compared to the nuclear arsenals we have today, their destructive power was tiny. The bombs dropped on Hiroshima and Nagasaki were atomic, relying on the fission of uranium or plutonium; in contrast, the first H-bomb, which utilised the energy released from the fusion of hydrogen isotopes into helium, was developed in 1952 and was five hundred times more powerful.[48] The largest bomb tested had an explosive yield of fifty million tonnes—over three thousand times that of the bomb dropped on Hiroshima.[49] In parallel, the global stockpile of nuclear weapons rose many thousandfold, from two in 1945 to just over forty thousand in 1967. The overall destructive power of explosive weapons therefore increased enormously over the course of just two decades, with the vast majority of those weapons built by the United States and the Soviet Union.[50]

It would be a mistake to infer that, because an all-out nuclear war never occurred, it was very unlikely to have occurred. Indeed, there were several close calls. During the Cuban Missile Crisis, John F. Kennedy put the chance of all-out nuclear war at "somewhere between one in three and even."[51] In 1979, US command centres detected a large number of incoming nuclear missiles, causing them to begin preparing for their own counterstrike. But when senior commanders checked the raw data to confirm the strike, they saw no evidence of incoming missiles. Upon further investigation, they realized a training tape designed to simulate a Soviet nuclear strike had been accidentally playing on the command centre screens. Just four years later, during a period of heightened tensions between the United States and Soviet Union, a similar false alarm took place in a Soviet command centre after a Soviet early-warning system detected five incoming nuclear missiles.[52] The officer on duty, Stanislav

Petrov, was sceptical that a US first strike would involve just five nuclear missiles, and he couldn't find evidence of the missile's vapor trails. Based on this alone, he reasoned that the warning system must have been mistaken and correctly reported the warning as a false alarm. If he had not, Soviet protocol was to launch a counterstrike, though it is unclear whether those higher in command would have believed that it was not a false alarm.

Thankfully, total US and Russian stockpiles have fallen by a factor of seven since their peak in 1986. But they are still very high, with 9,500 nuclear warheads remaining.[53] And compared to total defence budgets, the cost to manufacture new nuclear warheads is very small. If there were a reignition of serious military tensions between the United States and Russia, or new military tensions between other nuclear powers like the United States and China, or India and Pakistan, nuclear arsenals could grow significantly.[54]

An all-out nuclear war would potentially kill a much larger percentage of the world than any catastrophe we have seen. The direct death toll alone would be measured in the tens to hundreds of millions.[55] Even worse, some modelling suggests that such a war could result in a "nuclear winter": if soot from the burning cities were lofted high enough to reach the stratosphere, then global average temperatures would drop by eight degrees Celsius, returning to normal only over the course of ten to twenty years.[56] This would make it impossible to grow food across much of the Northern Hemisphere for several years, though agriculture would still be feasible across much of the tropics and the Southern Hemisphere, albeit hampered by reduced rainfall in many places.[57] Some argue that this could lead to widespread famine, potentially putting billions at risk of starvation.[58]

For concreteness, let's consider what I would regard as an absolute worst-case nuclear scenario, in which 99 percent of the world population dies in the aftermath of an all-out war, leaving a global population of around eighty million. This is perhaps possible if weapons stockpiles greatly expand and weapons become much more powerful, or if other weapons, such as bioweapons, are also used. Using my definition of civilisational collapse as an event in which society loses the ability to create most industrial and postindustrial technology, we can now try to answer the first question: If 99 percent of the population died, would civilisation collapse?

Up until recently, this question had only very limited investigation, so I commissioned a report on the topic from Luisa Rodriguez, a researcher for Rethink Priorities who subsequently came to join my team. Luisa does not fit the typical stereotype of a "prepper"—someone who worries about and prepares for societal catastrophe. The daughter of a socialist who fled El Salvador and gained asylum in the United States, for most of her life she worked on pretty typical issues for a socially conscious member of the Left: as a teenager, she wanted to be a Peace Corps volunteer like her grandparents, and during university she oscillated between pursuing a career as an infectious disease doctor and one in international development nonprofits. Now she possesses a small stash of survivalist tools: heirloom seeds, because many of the plants grown on modern farms are hybrids that do not guarantee that desirable traits will be passed on the next generation; a flint-based lighter, because making fire is difficult; and a hand-crank emergency generator. On a date night with her partner, they created a plan for what to do if an apocalypse occurred, including where to meet if all communications infrastructure was down. I found this strangely romantic.

For all this, Luisa is fairly optimistic about the robustness of civilisation in the face of catastrophe. I share this qualified optimism: society *probably* would not collapse. But it is difficult to be completely sure, and when the stakes are so high, the risk of nonrecovery should be taken very seriously.

One set of reasons for optimism comes from the examples of postcatastrophe societies we have just discussed, such as Europe after the Black Death, Hiroshima, and Cuba. Even in the face of enormous local catastrophes, society recovered remarkably quickly.

There are also specific reasons to think that civilisation would not collapse if 99 percent of people died. Much of the physical infrastructure like buildings, tools, and machines would be preserved and could be used after the catastrophe. Similarly, most knowledge would be preserved, in the minds of those still alive, in digital storage, and in libraries: there are 2.6 million libraries in the world, with hundreds of thousands in countries without either nuclear weapons or alliances with countries with nuclear weapons.[59] Critical skill sets would still remain: even if a catastrophe killed 99 percent of people, the chance that among the survivors there would be fewer than one hundred aeroplane engineers, nuclear power plant workers, organic chemists, or telecommunications engineers is close to zero. Two billion people today work in agriculture, with a sizable fraction working in smallholder subsistence farms, so it is exceptionally unlikely that we would lose all knowledge of agriculture.[60]

Finally, any large-scale catastrophe would be quite diverse in its impacts. Because all countries with nuclear weapons are in the Northern Hemisphere, the impacts of a nuclear winter would be more limited in the Southern Hemisphere; and because oceans retain heat, coastal areas would be much less affected.[61] For coastal South America or Australia, a nuclear winter would result in a summer about five degrees cooler than usual,[62] which would be bad but manageable. Similarly, if bioweapons were used, some island nations that were not involved in the conflict might be better able than other countries to defend against them by closing their borders. (Often, when worst-case disasters are modelled, New Zealand tends to come out relatively unscathed, which is why so many ultrarich preppers buy property there.[63]) So when we imagine a world in which 99 percent of people have died, we should not imagine this as being uniform across the world; rather, some countries would be devastated and some comparatively unaffected.

This makes the chance of global recovery higher. Those countries, perhaps Australia and New Zealand, that would not be directly affected would have their population, infrastructure, knowledge base, and political and civil institutions intact. And they could be self-sufficient: Australia and New Zealand already grow several times the amount of food required to sustain their own population; between them, they have ample fossil fuel reserves.[64] Even in the wake of such an unprecedented disaster, civilisation would continue.

As a sanity check on this argument, we could think about the last time that the world population was at eighty million people, which was very roughly in 2,500 BC.[65] At this time, although global civilisation was much less technologically sophisticated than today, it was not on the brink of collapse, and, on balance, I think a postcatastrophe world would be better off than the world in 2,500 BC because of the knowledge, physical capital, and institutions we have developed over the last 4,500 years.[66]

Now let's turn to the second question. Suppose that there were some catastrophe that resulted in the complete collapse of global civilisation, and we could rely only on preindustrial technology. Perhaps the considerations I've given in the previous paragraphs are mistaken in some way, and a war that killed 99 percent of people really would be sufficient for global civilisational collapse. Or perhaps some other, even larger catastrophe occurred, killing 99.999 percent of the world population, leaving only tens of thousands of people. If this happened, would we lose agriculture, and if we did, would we ever get it back? Or would we remain in hunter-gatherer or farming societies for millions of years, until some natural disaster like an asteroid strike killed us off?

In part for the reasons mentioned above, it is difficult to see why agriculture would stop after a collapse. If the world population shrank to eighty million, it is extremely likely that enough survivors would have knowledge of agriculture. The last time the world population was eighty million, in 2500 BC, we were already well into the agricultural revolution. Even if the global population fell to tens of thousands, it is still likely that some of the survivors would have knowledge of agriculture. Moreover, we would be in a much better position to maintain agriculture relative to people in 2500 BC. It took thousands of years for us to domesticate wild plants to make them better suited to farming, slowly (and mainly inadvertently) selecting those plants that bore the highest yields. The difference between modern domesticated plants and their wild ancestors is truly extraordinary. For example, the maize we eat today is around ten times larger than its wild ancestor, teosinte.[67] Likewise, the wild ancestor of watermelon was half the size, had pale white flesh, and was much less sweet than modern watermelons, while the wild ancestor of the modern tomato was only slightly larger than a pea.[68] Access to these domesticated plants would leave us in a much better place than early agriculturalists.

This does not mean that agricultural yields would immediately be as high as they are today.[69] High modern yields depend in large part on industrial products such as synthetic fertiliser, insecticides, and pesticides. Without these, many crops would be lost to weeds and pests. In addition, many domesticated plants are hybrids: they are produced by crossing two inbred strains to produce one high-yielding strain.[70] Hybrid crops lose their desirable properties over the generations. If there were a break in agriculture, some important varieties of some of our staple crops, in particular maize and to a lesser extent rice, would probably be lost.[71] However, many strains of our staple crops, including most strains of wheat and soybeans and many strains of rice, are not hybrids, so they would likely survive.[72]

Another key factor would be that, depending on the catastrophe, the longterm climatic conditions that seem to be necessary for agriculture would still be in place. Agriculture was developed at least ten times across history, at different times and in different places.[73] Archeobotanists have found evidence that societies in Mesopotamia domesticated wheat, barley, rye, and figs between 11,000 BC and 8000 BC. People in South and Central America independently domesticated squash at around the same time, in 8000 BC. Three thousand years later Papua New Guinea domesticated yams, bananas, and taro. This happened again and again, among societies that never crossed paths, with entirely different crops, thousands of years apart.[74] This happened as we transitioned out of the last ice age into the warmer period that we still live in today, known as the Holocene.

The reason the Holocene has been conducive to agriculture is that it is warm, so frost does not destroy the growing season; it has higher carbon dioxide levels, which is good for crop yields; and it is climatically stable.[75] If there were a collapse, we would, due to climate change, probably live in an environment one to three degrees warmer than today's. But this seems unlikely to make a major difference: generally it is cold and low–carbon dioxide environments that make global agriculture near impossible, not warm and high–carbon dioxide environments.

So it seems very likely that agriculture would survive a catastrophe or would be quickly redeveloped, even if the total human population dropped to as few as tens of thousands of people. So, assuming that agriculture survived, would we reindustrialise? Unlike the development of agriculture, the Industrial Revolution happened only once; perhaps the conditions that gave rise to it were therefore highly contingent. However, there are a few reasons for thinking that industrialisation is probably not a bottleneck either.

First, it took only around thirteen thousand years for the Industrial Revolution to occur after the very first development of agriculture; if industrialisation were an incredibly unlikely event, we would expect it to have taken much longer.[76] Of course, thirteen thousand years is a long time from the perspective of a single human life, but it's a short time from the perspective of a species: given the typical life spans of other mammals or hominins, even after a major catastrophe we would still have many hundreds of thousands of years ahead of us.

The second reason for thinking that we'd reindustrialise after civilisational collapse is that the generations following a global catastrophe would in some ways have a serious head start over our predecessors. Some stone and concrete buildings would last hundreds of years.[77] While most tools and machines would degrade within a few decades, some would be preserved in modern buildings and would be functional.[78] Even if only a tiny fraction of tools and machines survived, this would ensure that the postcollapse survivors would know that such technology was possible, and they could reverse engineer some of the tools and machines that they found. Knowledge of industrial technology would be preserved in libraries, as would knowledge of politics and economics, which would allow embryonic states to copy successful policies.

Indeed, there is evidence that industrialisation happens fairly quickly (on historical timescales) once the knowledge of how to industrialise is there. Once Britain industrialised, other European countries and Western offshoots like the United States quickly followed suit; it took less than two hundred years for most of the rest of the world to do the same. This suggests that the path to rapid industrialisation is generally attainable for agricultural societies once the knowledge is there.

A final reason for thinking we'd reindustrialise is that there would be strong incentives for postcollapse societies to do so, such as improving living standards or gaining power over local competitors.

Climate Change

So far, I have looked at catastrophe as a result of war or accidental release of engineered pathogens. But what about climate change—could it cause global civilisation to collapse?

One cause for optimism is that we are making real progress on climate change: recent years have given us more cause for hope than any other point in my lifetime.[79] The International Energy Agency predicts that global coal use peaked in 2014 and is now in structural decline.[80] The main reason for the decline in coal use to date is competition from cheap natural gas,[81] but a more fundamental future shift is now under way. This is in significant part due to environmental activism, which has changed the climate prognosis in two ways.

First, thanks in part to youth activism, attention towards climate change has increased significantly, and several key players have made ambitious climate pledges, most notably China, which plans to reach zero emissions by 2060, and the European Union, which is aiming for 2050; and efforts are increasing at the state level in the United States.[82]

Second, there has also been huge progress on key low-carbon energy technologies: solar, wind, and batteries.

Thanks to long-standing policy support from environmentally motivated governments, the cost of solar panels has fallen by a factor of 250 since 1976, while the cost of lithium ion batteries has fallen by a factor of 41 since 1991.[83] Even though solar and wind supply only around 3 percent of energy today, if the exponential cost declines continue, in twenty years they will supply a substantial fraction of global energy.[84] Similarly, in the next few years, the total cost of ownership for electric cars—including purchase, fuel, and maintenance costs—is projected to drop below that of petrol and diesel cars.[85]

Figure 6.3. Global average price of solar photovoltaic (PV) modules, measured in 2019 US$ per watt (i.e., adjusted for inflation).

However, we shouldn't get complacent. There is a substantial chance that our decarbonisation efforts will get stuck. First, limited progress on decarbonisation is exacerbated by the risk of a breakdown in international coordination, which could happen because of rising military tensions between the major economies in the world, which I discussed in Chapter 5. Decarbonisation is a truly global problem: even if most regions stop emitting, emissions could continue for a long time if one region decides not to cooperate. Second, the risk of prolonged technological stagnation, which I discuss in the next chapter, would increase the risk that we do not develop the technology needed to fully decarbonise. These are not outlandish risks; I would put both risks at around one in three.

For the purposes of assessing civilisational collapse, let's ask about the low-probability but worst-case climate scenario, in which we ultimately burn through all recoverable fossil fuels. (In higher-end estimates, these amount to three trillion tonnes of carbon,[86] so if our emissions remain at current levels, this would take about three hundred years.) If we did so, there would most likely be around 7 degrees of warming relative to the preindustrial period, and a one in six chance of 9.5 degrees of warming.[87]

The effect of such extreme climate change is difficult to predict. We just do not know what the world would be like if it were more than seven degrees warmer; most research has focused on the impact of less than five degrees.[88] Warming of seven to ten degrees would do enormous harm to countries in the tropics, with many poor agrarian countries being hit by severe heat stress and drought.[89] Since these countries have contributed the least to climate change, this would be a colossal injustice.

But it's hard to see how even this could lead directly to civilisational collapse. For example, one pressing concern about climate change is the effect it might have on agriculture. Although climate change would be bad for agriculture in the tropics, there is scope for adaptation, temperate regions would not be as badly damaged, and frozen land would be freed up at higher latitudes.[90] There is a similar picture for heat stress. Outdoor labour would become increasingly difficult in the tropics because of heat stress, which would be disastrous for hotter and poorer countries with limited adaptive capacity. But richer countries would be able to adapt, and temperate regions would emerge relatively unscathed.[91]

What about feedback loops, where some amount of warming leads to further warming? Two possibilities that have been raised are "moist greenhouse" and "runaway greenhouse" effects. In both scenarios, temperatures become so hot that the oceans are lost to space, as has occurred on Venus. But the existing models suggest that it is not possible to trigger a runaway greenhouse on Earth by burning fossil fuels.[92] It also seems unlikely that we could trigger a moist greenhouse, but if carbon dioxide did cause a transition to a moist greenhouse state, carbon dioxide concentrations would naturally decline over hundreds of thousands of years, well before the earth's water would be lost to space.[93]

There are other possible feedback effects that look more concerning. In what is probably the most alarming climate science paper in recent years, one model found that once carbon dioxide concentrations reach around 1,300 parts per million, stratocumulus clouds will burn off and there will be eight degrees of warming over the course of years, on top of the six to seven degrees we will already have lived through.[94] If we burned three trillion tonnes of carbon, atmospheric carbon dioxide concentrations would reach around 1,600 parts per million, so this threshold is within reach.[95]

This research is controversial, and scientists are divided on how plausible it is.[96] Unfortunately, it is just difficult to know how great the risk of this kind of feedback is because carbon dioxide concentrations have not been greater than 1,300 parts per million for at least tens of millions of years.[97] But even a low probability that there could be feedback effects of this sort should greatly concern us. It is hard to know what the impact of eight degrees of warming over a few years would be, and this question has not been researched by the scientific community. Climatic instability is generally bad for agriculture, although my best guess is that global agriculture would still be possible even during this extreme transition: even with fifteen degrees of warming, the heat would not pass lethal limits for crops in most regions.[98] But it is hard to know exactly what would happen because such a change would be so extreme and so unprecedented. Possible nonlinear tipping points like this are, in my view, the greatest threat that climate change poses to our longterm future.[99]

Even if climate change does not drastically increase the risk of civilisational collapse, it might well make it harder to recover from collapse caused by some other event, like a nuclear or biological war. For the reasons mentioned above, it seems that agriculture would still be possible even if there were high levels of warming. But it would mean that industrial civilisation would have to reemerge in a warmer world than we faced historically, which should increase our uncertainty about our prospects for recovery.

Importantly, climate change lasts for a very long time: temperatures would be similar after ten thousand years and would only return to normal after hundreds of thousands of years.[100] The sheer length of time before temperatures would return to current levels is long enough that, if climate change does delay recovery, almost all machines, tools, and buildings will have degraded; almost all books in libraries will have decayed; and knowledge passed down from one generation to another may have progressively gotten corrupted.[101]

Fossil Fuel Depletion

Burning fossil fuels produces a warmer world, which may make civilisational recovery more difficult. But it also might make civilisational recovery more difficult simply by using up a nonrenewable resource that, historically, seemed to be a critical fuel for industrialisation. Our preindustrial ancestors primarily relied on animal and human muscle, and on the burning of biomass such as wood or crops. This all changed at the start of the Industrial Revolution, which marked the beginning of centuries of almost-unchecked fossil fuel burning. On the path to industrialisation and out of poverty, countries begin by burning prodigious amounts of fossil fuels, usually, though not always, starting with coal and then shifting to oil and gas.[102]

Since, historically, the use of fossil fuels is almost an iron law of industrialisation, it is plausible that the depletion of fossil fuels could hobble our attempts to recover from collapse. Although countries have so far almost always industrialised with fossil fuels, would that have to be true in a postcollapse world? If we have run out of coal, oil, and gas, why could we not have a green industrial revolution instead? This question has received relatively little attention, and I am only aware of one sophisticated discussion of it, by Lewis Dartnell, who has spent the last few years researching how we might bounce back from catastrophe.[103]

If civilisation collapsed, we might be able to get some electricity out of some of the remaining solar and wind farms. However, this would not last long. Solar panels and wind turbines degrade over the course of a few decades. It would be fiendishly difficult to create them from scratch once advanced international supply chains, such as the silicon purification factories necessary for solar panels, have been destroyed. Solar and wind also could not provide the high-temperature heat that is necessary for several crucial industries, such as cement, steel, brick, and glass.[104] In a postcollapse world, it would be very difficult to mine and transport nuclear fuel and to power up, run, and maintain technologically complex nuclear-power stations. So nuclear-powered reindustrialisation seems unlikely.

An alternative fuel is charcoal. Charcoal is wood that has been pyrolyzed: heated without oxygen in order to remove water. It has roughly the same energy density of coal, can substitute for it, and is renewable. Brazil's steel industry, which is the ninth largest in the world, relies on charcoal to produce high-temperature heat. So we know that charcoal can power some advanced industries. The problem is that it's not clear whether we would be able to redevelop the efficient steam turbines and internal combustion engines needed to harness the energy from charcoal. In the Industrial Revolution, steam turbines were first used to pump out coal mines to extract more coal. As Lewis Dartnell says, "Steam engines were themselves employed at machine shops to construct yet more steam engines. It was only once steam engines were being built and operated that subsequent engineers were able to devise ways to increase their efficiency and shrink fuel demands. They found ways to reduce their size and weight, adapting them for applications in transport or factory machinery. In other words, there was a positive feedback loop at the very core of the industrial revolution: the production of coal, iron and steam engines were all mutually supportive."[105]

It took a lot of easily accessible energy to develop the technologies required for the Industrial Revolution. To do the same again, we would need an enormous amount of wood, which would require a lot of land. This would compete with agriculture, which would be straining to feed a growing population.

After assessing the prospects of a postcollapse recovery, Lewis Dartnell concluded that an industrial revolution without coal would be, at a minimum, very difficult. This consideration could be of major importance. If a catastrophe that falls short of killing us all but causes us to lose industrial technology is many times as likely as a catastrophe that causes human extinction, and if the depletion of easily accessible fossil fuels makes recovery from such a catastrophe many percentage points less likely, then the depletion of fossil fuels could contribute a similar amount to the risk of the end of civilisation as the risk of human extinction.

If fossil fuels are potentially so important to reindustrialisation, we should ask: How much do we have left? There are about twelve trillion tonnes of carbon remaining in fossil fuel resources, of which 93 percent is coal. However, only a fraction of the fossil fuels are ultimately recoverable, and a much smaller fraction are easy to access.[106] Data on global surface coal reserves are surprisingly limited, but one study from 2010 found that there are two hundred billion tonnes of carbon remaining in surface coal.[107]

Easy-to-access coal would be especially important in a postcollapse world in which we have regressed to preindustrial technology. Some surface coal can be accessed with minimal digging and can be recovered using technology as simple as a shovel. Western Europe has already burned through almost all its easy-to-access coal. Most easy-to-access coal is now in China, the United States, India, Russia, and Australia.[108] The North Antelope Rochelle coal mine in the United States (the largest coal mine in the world) contains nine hundred million tonnes of carbon in easy-to-access recoverable coal.[109] This single mine alone could fuel the first few decades of reindustrialisation.[110] The amount of surface coal remaining worldwide would be enough to provide all of the energy we used between 1800 and 1980.[111]

However, these resources may not be around forever. If surface coal production stays constant, recoverable surface coal will last for more than three hundred years in the United States, for more than two hundred years in Russia and China, and for fifty to one hundred years in India and Australia.[112] At present, demand for coal is falling globally and environmental regulations are being strengthened, so surface coal will probably last longer than this.[113] But from a longterm point of view, we need to take these sorts of timescales seriously. The more we deplete these resources, the more we imperil our chances of reindustrialisation.

How likely is it that we will burn through these reserves? I see three ways this could happen. First, civilisational collapse would mean that, in the course of returning to modern levels of technology, we would probably burn through almost all remaining easy-to-access fossil fuels. Even if we have enough reserves to recover from civilisational collapse once, we wouldn't have enough if civilisation collapsed a second time. This might not be as unlikely as it seems: if civilisation has collapsed once, that suggests that civilisational collapse is not extremely unlikely, and it might well happen again.[114]

Second, we might fail on the "last mile" of decarbonisation—eliminating the hardest-to-replace quarter of emissions, such as the use of coal to provide high-temperature heat in the cement and steel industries.[115] To wholly do away with fossil fuels, we'll need a suitable combination of cheap, controllable low-carbon power and cheap zero-carbon fuels such as hydrogen. While innovative ways to improve these capabilities have been proposed, it is unclear whether we will get there.[116]

Worse, solving decarbonisation through the wrong mix of technologies might backfire: the final way we might continue to burn a lot of fossil fuels is if we make extensive use of carbon capture and storage. Carbon capture and storage involves capturing carbon at point sources such as power plants and then burying it underground. Carbon can also be captured from the ambient air in a process known as "negative emissions."

Carbon capture would remove a large fraction of the environmental costs of fossil fuels (though the terrible air pollution costs would remain). Consequently, carbon capture would weaken the reason for environmentally motivated governments to stop burning fossil fuels in the first place. This is great insofar as it reduces damages from climate change. But it could significantly increase the risk that we keep burning fossil fuels indefinitely, using up the easily accessible resources and undermining the prospects for recovery in the event of civilisational collapse.

All in all, my best guess is that we will phase out most fossil fuel burning this century. However, depending on what happens with relevant technological progress, I still think there is a significant chance that we will continue to burn coal and other fossil fuels for a long time. If so, we would use up a resource that might be crucial for recovery after the collapse of civilisation.

Conclusion

An all-out nuclear war, perhaps supplemented by bioweapons, would be utterly devastating. Yet the risks from weapons of mass destruction and a potential war between the world's major powers have largely fallen out of the mainstream conversation among those fighting for a better world. I find this both striking and concerning. Although such a catastrophe is, in my view, unlikely to lead to unrecovered civilisational collapse, it is difficult to be extremely confident that it won't. This lingering uncertainty is more than enough to make the risk of unrecovered collapse a key longtermist priority.

This risk is exacerbated considerably by our continued burning of fossil fuels. If we fail to wholly decarbonise and burn through the easily accessible fossil fuels, then the odds that we will be able to bounce back from civilisational collapse get much worse.

The chance of the end of civilisation this century, whether via extinction or permanent collapse, is far too high for us to be comfortable with. In my view, giving this a probability of at least 1 percent seems reasonable. But even if you think it is only a one-in-a-thousand chance, the risk to humanity this century is still ten times higher than the risk of your dying this year in a car crash.[117] If humanity is like a teenager, then she is one who speeds round blind corners, drunk, without wearing a seat belt.

And that is just for the risk this century. If we want humanity to survive and flourish over the long term, we need to both make catastrophic risks as small as possible and ensure they stay small indefinitely. But if society stagnates technologically, it could remain stuck in a period of high catastrophic risk for such a long time that extinction or collapse would be all but inevitable. I turn to this possibility in the next chapter.

CHAPTER 7

Stagnation

Efflorescences

In the eleventh century, the world's epicentre of scientific progress was Baghdad, during an era known as the Islamic Golden Age.[1] This era produced an astonishing assortment of discoveries and innovations: we understood for the first time how magnifying lenses work, invented a flywheel-powered water-lifting device, built the earliest programmable machine (a flute-playing automaton), and discovered the first code-breaking method.[2] The words "algorithm" and "algebra" both come from Arabic, and even the Hindu-Arabic number system we use (1, 2, 3, etc.), was imported into Europe in the thirteenth century by Fibonacci, who had travelled throughout the Mediterranean world to study under the leading Arabic mathematicians of the time.[3] Translated scientific works from the medieval Islamic world are believed to have played a central role in fuelling the Renaissance and the Scientific Revolution in Europe.[4]

However, the Islamic Golden Age did not last: from the twelfth century AD onwards, the rate of scientific progress slowed considerably.[5] There are a number of explanations for why this occurred. Some point to the Mongol invasion; others to the role of the Crusades; others to a cultural shift that encouraged theological work over scientific inquiry.[6]

The Islamic Golden Age is one example of what historian Jack A. Goldstone calls an *efflorescence*: a short-lived period of technological or economic advancement in a single culture or country.[7] There have been many efflorescences throughout history. Ancient Greece may be another example. From 800 to 300 BC, living standards improved substantially, as did life expectancy; the typical Greek house grew from roughly 80 square metres (about 860 square feet) to 360 square metres (about 3900 square feet) and became much better built.[8] This economic progress coincided with an extraordinary flourishing of intellectual progress: we still read Plato, Aristotle, Herodotus, Thucydides, and many more ancient Greek writers today.

What is different about the modern growth era is that technological progress and economic growth have been sustained to reach much greater heights. With the Industrial Revolution, the world moved to unprecedentedly rapid rates of growth and technological progress, which continue to this day.

But will this continue? In Chapter 4, we saw that there was a case for thinking that, by automating the process of technological innovation, artificial intelligence could bring about even faster technological progress than we've seen to date. In this chapter we'll consider the opposite possibility. Perhaps future historians will look back on our era just as a really big efflorescence that, like other efflorescences before us, was followed by stagnation. My concern here is not just with a slowdown in innovation but with a near halt to growth and a plateauing of technological advancement.

Though indefinite stagnation seems unlikely to me, it seems entirely plausible that we could stagnate for hundreds or thousands of years—a sort of civilisational interregnum. That would be of great longtermist importance for two reasons. First, the society that emerges from the interregnum might be guided by very different values than society today. Second, and more clearly, a period of stagnation could increase the risks of extinction and permanent collapse.

To see this second point, consider what would have happened if we had plateaued at 1920s technology. We would have been stuck relying on fossil fuels. Without innovations in green technology, we would have kept emitting an enormous amount of carbon dioxide. Not only would we have been unable to stop climate change, but we would also have simply run out of coal, oil, and gas eventually. The 1920s' level of technological advancement was *unsustainable*. It's only with the technological progress of the last hundred years that we have the capability to transition away from fossil fuels.

Our next level of technological advancement might be unsustainable, too. We could face easy-to-manufacture pathogens and other potent means of destruction without sufficient technology to defend against them. There would be a constant risk of a civilisation-ending catastrophe. If we stayed stuck at this unsustainable level for long enough, such a catastrophe would be essentially inevitable. To safeguard civilisation, we therefore need to make sure we get beyond that unsustainable level and reach a point where we have the technology to effectively defend against such catastrophic risks.

The idea of sustainability is often associated with trying to slow down economic growth. But if a given *level* of technological advancement is unsustainable, then that is not an option. We may be like a climber scaling a sheer cliff face with no ropes or harness, with a significant risk of falling. In such a situation, staying still is no solution; that would just wear us out, and we would fall eventually. Instead, we need to keep on climbing: only once we have reached the summit will we be safe.[9]

Is Technological Progress Slowing Down?

The economic data suggest that technological progress is already slowing down. To measure the rate of technological progress, we can look at what economists call "total factor productivity." Though this term is complex, the idea is simple. There are two ways by which economic output could increase. First, inputs could increase: there could be more people working, or people could buy and use more machines, or they could use more natural resources. Second, we could increase our ability to get more output from the same inputs. Total factor productivity measures this ability and represents technological advancement. To illustrate, think about how many calories of food you can produce from an acre of land (a fixed input): because of fertilisers and modern farming techniques, we now produce far more than farmers throughout history could have done, and farmers historically produced far more than hunter-gatherers could.

When economists have measured this, they've found that the growth rate of total factor productivity in the United States has been generally declining over the last fifty years.[10]

Qualitatively, too, it seems that rates of technological progress have slowed down. To see this, consider a thought experiment from the economic historian Robert Gordon.

Imagine you are a typical inhabitant of the United States in 1870.[11] You live on a rural farm; you produce most of your food and clothing yourself.

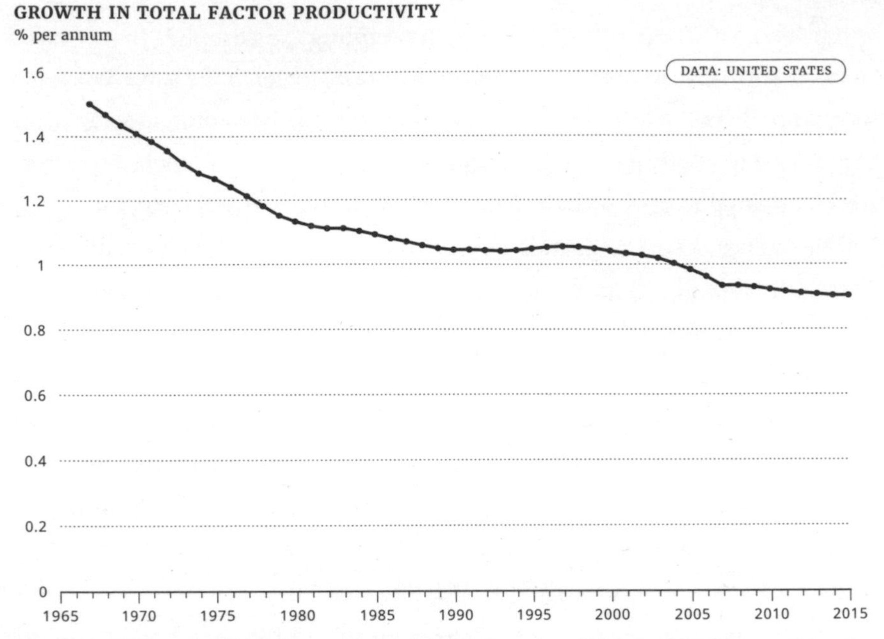

Figure 7.1. Smoothed trend of US quarterly total factor productivity (TFP) data. Growth in TFP in the United States has been declining over the last fifty years.

Your only sources of light are candles, whale oil, and gas lamps if you're lucky. If you're a man, you face gruelling physical labour, sometimes from the age of twelve onwards. If you're a woman, you face unrelenting toil as a housewife: one calculation found that in 1886 "a typical North Carolina housewife had to carry water 8 to 10 times a day. . . . Over the course of a year she walked 148 miles toting water."[12] You rely on horses for transport. Mostly your life is one of isolation: the telephone doesn't yet exist, and the postal service doesn't reach your farm. Life expectancy at birth is thirty-nine years,[13] and modern forms of leisure are unknown. The tallest building in New York City is a church steeple.

Now, suppose that one morning, you wake up and it's fifty years later, the year 1920. Your standard of living is in the process of rapid and dramatic improvement. The electrification of America is well underway, reaching close to half of American households. If you are lucky enough to have electricity, the lighting it provides is ten times brighter than the kerosene lamps that preceded it and a hundred times brighter than the candles that preceded those. People are beginning to use telephones, which enable instant communication. Mass-produced cars are beginning to replace horses, with nearly a third of the population owning a car. Life expectancy is now sixteen years greater, at fifty-five years. You are less likely to contract cholera or typhoid thanks to routine disinfection of drinking water. Skyscrapers are beginning to rise in New York City.

Next, suppose you wake up fifty years later again, in 1970. As a typical US inhabitant, you again see an enormous difference in your life. Most households finally have an indoor flush toilet. You live in a spacious suburban home with a gas stove, a refrigerator, and central heating. Your household owns two cars, and if you want you can fly around the world on an aeroplane. You have a television, and on this TV you just watched a man land on the moon. You have penicillin and new vaccines, such as against polio; life expectancy is sixteen years longer again, at seventy-one. Your work is probably much less exhausting, and with a forty-hour workweek, vacations, and retirement, you have ample leisure time.

Finally, imagine waking up fifty years later again, in 2020. Comparatively speaking, this time your life is not all that different. Among your household appliances, the only difference is that you now have a microwave. Your television is bigger and higher definition, and you have a wider range of shows to watch. You still use cars to get around, though they are now safer and easier to drive. Life expectancy has increased but more moderately, by only eight years, to seventy-nine years. Of course, there has been a revolution in information and communication technologies—you now have computers and the internet, tablets and mobile phones. But technological progress that meaningfully impacts your life has been confined nearly exclusively to those spheres.

From 1870 to 1970, there were extraordinary advances made in a wide number of different industries. This included information and communication technologies such as the telephone, radio, and television, but it also included advances in many other industries, such as transportation, energy, housing, and medicine. Since 1970, there's been substantial progress in information and communication technologies, but in all those other industries, progress has been comparatively incremental. Since 1970, the pace of progress seems to have slowed.

The economist Tyler Cowen has argued that a growth slowdown is extremely bad from a longterm perspective.[14] Decreases to the rate of economic

Table 7.1. Assorted Changes in the Standard of Living in the United States

	1870	1920	1970	2020
Income per capita (in 2011 dollars)	$4,800	$10,200	$24,000	$55,300
Life expectancy (in years)	39	55	71	79
Height of the tallest building in New York City (in feet)	281	792	1,472	1,776
Transcontinental journey time	Wagon: more than 5 months Stagecoach: more than 25 days Transcontinental railroad (completed 1869): 6 days	Railroad: 3 days	Jet airplane: half a day	Jet airplane: half a day
Percent of households with running water	<20%	~55%	98%	>99%
Percent of families with electric lighting	0%	35%	99%	>99%
Communication	Postal service, telegraph (only 5% of towns)	Telephone in 35% of households	Telephone in 90% of households, and much cheaper	Cell phones, internet
Entertainment and information	Newspapers	Cinema (still silent). Radio later in the 1920s.	TV	Internet
Annual working hours per worker	3,100 (~60 hours a week)	2,500	1,900 (~40 hours a week)	1,750

Note: For data sources, see whatweowethefuture.com/notes.

growth, he argues, would be hugely harmful to future generations. For example, suppose that the long-run growth rate slows from 2 percent per year to 1.5 percent per year. The difference this makes for people in a hundred years' time will be massive: they will be nearly 40 percent poorer at a 1.5 percent growth rate than they would have been at a 2 percent growth rate.

However, from a truly longterm perspective—thinking in terms of thousands or millions of years or more—this argument loses force, simply because exponential economic growth can't go on forever. As I suggested in Chapter 1, if current growth rates continued for just ten thousand years, then we would have to start producing trillions of present-civilisations' worth of output for every atom within reach. But this seems unlikely to be possible. At some point, economic growth must plateau.

But if so, then speeding up or slowing down the world's economic growth rate is not making a contingent change to civilisation's long-run trajectory. To illustrate, suppose that at a long-run growth rate of 2 percent per year, we would reach the plateau of economic growth in 1,000 years. If instead we go through a century of slower growth, at only 1.5 percent annually, we would reach that economic plateau in 1,025 years instead.[15] The world would be poorer than it otherwise would have been for 1,025 years, but our destination would be the same, and there would be no difference to the world in economic output in all the time that followed.

A mere slowdown in technological progress would probably not make an enormous difference to the long-run trajectory of civilisation. But a period

THREE WAYS TO IMPROVE THE FUTURE

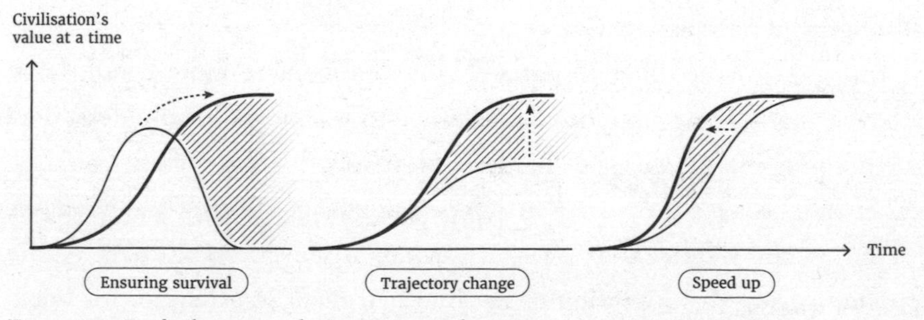

Figure 7.2. In the long run, the importance of ensuring survival and improving our trajectory dwarfs the importance of accelerating progress, assuming that acceleration doesn't change the longterm state we end up in.

of *stagnation*, where there is almost no progress at all for centuries or millennia, could be a much bigger deal.

How Likely Is Stagnation?

When economists discuss economic growth, they usually consider timescales of a few decades at most. We are interested in longer timescales—and there we are confronted with a vast range of possibilities. Simply extrapolating the trends of the last hundred years may not be very sensible. Just as growth in the year 2000 was very different from growth in 1700, growth in 2300 could look very different from growth today. There are a few growth economists, like Stanford professor Chad Jones, who have done pioneering work considering longer timescales.[16] In their models, both faster-than-exponential growth and near-zero growth arise quite naturally and should be taken seriously as possibilities.[17]

Why would growth decline to near zero? In brief, the argument goes as follows. Economists almost universally agree that in the long run, economic growth is driven by technological progress.[18] But as we make technological progress, we pick the low-hanging fruit, and further progress inherently becomes harder and harder. So far, we've dealt with that by throwing more and more people at the problem. Compared to a few centuries ago, there are many, many, many more researchers, engineers, and inventors. But this trend is set to end: we simply can't keep increasing the share of the labour force put towards research and development, and the size of the global labour force is projected to peak and then start exponentially declining by the end of this century.[19] In this situation, our best models of economic growth predict the pace of innovation will fall to zero and the level of technological advancement will plateau.[20]

Let's look at the different parts of this argument in more detail. First, after we make some amount of scientific and technological progress, does further progress get easier or harder? Intuitively, it seems like it could go either way because there are two competing effects. On the one hand, we "stand on the shoulders of giants": previous discoveries can make future progress easier. The invention of the internet made researching this book, for example, much easier than it would have been in the past. On the other hand, we "pick the low-hanging fruit": we make the easy discoveries first, so those that remain are more difficult. You can only invent the wheel once, and once you have, it's harder to find a similarly important invention.

Though both of these effects are important, when we look at the data it's the latter effect, "picking the low-hanging fruit," that predominates. Overall, past progress makes future progress harder.

It's easy to see this qualitatively by looking at the history of innovation. Consider physics. In 1905, his "miracle year," Albert Einstein revolutionized physics, describing the photoelectric effect, Brownian motion, the theory of special relativity, and his famous equation, $E=mc^2$. He was twenty-six at the time and did all this while working as a patent clerk. Compared to Einstein's day, progress in physics is now much harder to achieve. The Large Hadron Collider cost about $5 billion, and thousands of people were involved in its design, construction, and operation.[21] It enabled us to discover the Higgs boson—a worthy discovery for sure, but a small and incremental one compared to Einstein's contributions.[22]

In a recent article called "Are Ideas Getting Harder to Find?," economists from Stanford and LSE analysed this phenomenon quantitatively.[23] Across a range of industries, across firms, and in the aggregate economic data they found the same thing: progress becomes harder and harder. Based on their numbers, in order to double our overall level of technological advancement, we need to put in, conservatively, four times as much research effort as we did for the previous doubling.[24] To illustrate, suppose (simplistically) that initially it took 10 person-years of "research" to double the world's level of technological advancement: to move from knowing only how to make a stone axe to knowing how to make both an axe and a spear.[25] In order to get the next doubling of technological progress, it would take 40 person-years of research. The next doubling would take 160 person-years, then 640 person-years, then 2,560 person-years, and so on.

Some argue that this data on ideas getting harder to find simply reflects scientific institutions becoming more bureaucratic and less efficient. But the magnitudes are just too large. It's implausible that scientific institutions have become more than forty times less efficient since the 1930s, or more than five hundred times less efficient since 1800—which is what you'd need to believe to explain the data this way.[26] Rather, it's likely that additional progress inherently becomes harder the more progress one has already made.

Over the past century, we've seen relatively steady, though slowing, technological progress. Sustaining this progress is the result of a balancing act: every year, further progress gets harder, but every year we exponentially increase the number of researchers and engineers. For instance, in the United States, research effort is over twenty times higher today than in the 1930s.[27] The number of scientists in the world is doubling every couple of decades, such that at least three-quarters of all scientists who have ever lived are alive today.[28] So far, exponential growth in the number of researchers has compensated for progress becoming harder over time.

So to think about whether we can sustain technological progress, we have to think about whether we can keep exponentially growing the number of researchers. Consider that there are two ways to do this. First, you can increase the share of the population that is devoted to research. Indeed, we've been doing a lot of that, so that's been the source of most of US technological progress in the last few decades. Technology-driven growth of US per-capita incomes has averaged about 1.3 percent per year. A full percentage point of that comes from increasing the fraction of the population doing R&D and from improving the allocation of talent, such as by reducing gender and racial discrimination.[29]

The second way by which you can increase the number of researchers is by increasing the total size of the labour force: that is, you can grow the population. Over the last few decades, population growth has contributed about 0.3 percentage points to the United States' technologically driven per-capita growth rate.[30]

Historically, increasing population sizes have been a major factor in rates of technological progress. As Nobel Prize–winning economist Michael Kremer has noted, sheer population size seems to explain a big part of the very long-run comparative development of different geographic regions. With the end of the most recent ice age in 10,000 BC, five regions of the world became mutually isolated from one another: the Eurasian and African continents, the Americas, Australia, Tasmania, and Flinders Island.[31] By AD 1500, they had dramatically diverged technologically. The more populous a region was in 10,000 BC, the more complex their technology was by AD 1500. Eurasia had the most complex technology; the Americas followed, with cities, agriculture, and the Aztec and Mayan civilisations; Australia was in an intermediate position; while Tasmania had seen little technological development, and the population of Flinders Island had died out completely.[32] The larger the population, the more opportunities there were for people to invent new tools and techniques—more minds meant more inventions. And once a tool had been invented, that innovation would spread far and wide.

One effect of new technologies was that people could produce more calories from an acre of land. This enabled more people to live in a given region, which meant even more opportunities to invent new tools and techniques, which enabled a yet larger population—a feedback loop. Over time, this resulted in incredible growth in world population: from just a few million in 10,000 BC, to a few hundred million in AD 1, to one billion in 1800, to nearly eight billion today.[33]

For a long time, we saw a gradual accumulation of technology and population via this feedback loop. Technological progress took off in a particularly explosive way during and after the Industrial Revolution because we started dedicating a much greater fraction of society's efforts to science and technology.[34]

But we should not expect either of the two aforementioned trends—an ever-increasing population, of which an ever-increasing fraction is dedicated to research—to continue. The latter trend *cannot* continue indefinitely for the simple reason that at most 100 percent of the population can work in research. Right now, roughly 5 percent of US GDP is dedicated to R&D.[35] Maybe that can go to 20 percent, or maybe even higher, but we'd reach the practical limit well before the theoretical maximum of 100 percent.

The trend of an ever-growing population seems set to stall, too. The UN says world population will plateau by 2100, and researchers at the University of Washington predict an even earlier peak and subsequent decline.[36] That's because fertility rates are falling precipitously all around the world (see Figure 7.3). As people grow wealthier, they are choosing to have fewer children (see Figure 7.4).[37] This has been going on in rich countries for a while. The fertility rate is currently 1.5 children per woman in Germany, 1.4 children per woman in Japan, and 1.7 children per woman in the United States, in China, and in high-income countries on average.[38] As a result, the working-age population is now starting to peak and decline in these countries.[39] Much the same is true in poorer countries. South America's fertility

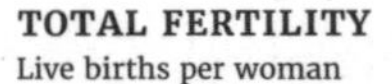

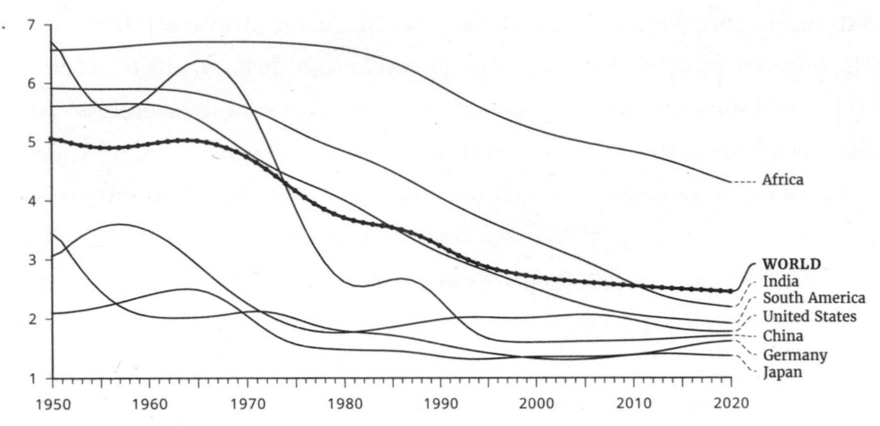

Figure 7.3. People have been having ever fewer children all over the world.

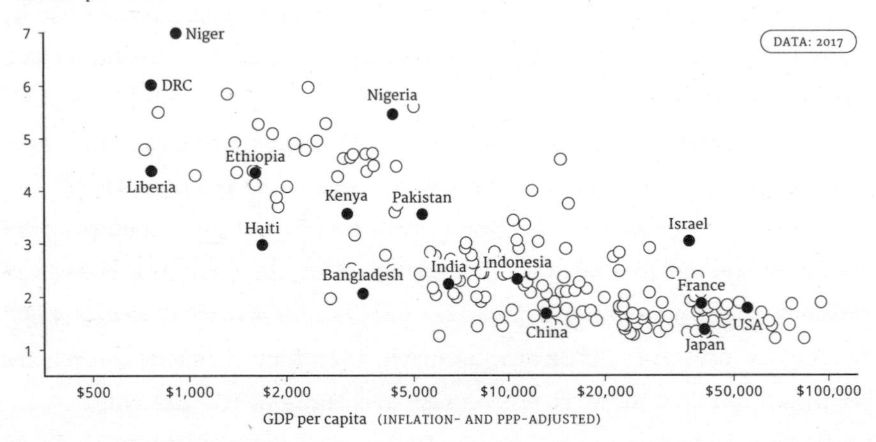

Figure 7.4. Children per woman against per-capita income (adjusted for price differences between countries); data for 2017.

rate is now just below 2, while India's fertility rate is at 2.2.[40] Africa is the only major continent expected to still have significant population growth over this century—but as African countries grow richer, their fertility rates are likely to drop, just like everywhere else.[41]

It's not just that world population will stop growing. Rather, the world might well be headed for an exponentially declining population.[42] As fertility rates are dropping everywhere, they aren't stopping at replacement rates—a bit above two children per woman—but are falling even lower, below replacement.[43] For twenty-three countries, including Thailand, Spain, and Japan, populations are projected to more than halve by 2100; China's population is projected to decline to 730 million over that time, down from over 1.4 billion currently.[44] Instead of ever more people, as we have had historically, we will have ever fewer people.

Think of the innovation happening today in a single, small country—say, Switzerland. If the world's only new technologies were whatever came out of Switzerland, we would be moving at a glacial pace. But in a future with a shrinking population—and with progress being even harder than it is today because we will have picked more of the low-hanging fruit—the entire world will be in the position of Switzerland. Even if we have great scientific institutions and a large proportion of the population works in research, we simply won't be able to drive much progress.

An increasing number of researchers and engineers from lower-income but high-growth countries and an increasing fraction of the population doing R&D in high-income countries could potentially increase the number of researchers and engineers by a factor of twelve or so.[45] That could be enough for another century's worth of technological progress. But thereafter, technological progress and economic growth will come to a near standstill.

You might think that, in the face of slowing technological progress, governments would step in to fix things. But this seems hard to do. First, they could try to get more people to work on R&D, for instance by increasing funding for universities. You might be able to make some gains by improving the efficiency of national grant-making bodies and other scientific institutions. But recall that every doubling of technological advancement takes roughly four times more research effort, so mere reductions of bureaucracy will only get you so far before almost the entire population is working in research.

Governments could try to increase the size of the labour force by making it more attractive for people to have kids. But the data suggest this is very hard to do. Many European countries have extensive child benefits, but their fertility rates tend to be even lower than in the United States. The Hungarian government has been spending up to 5 percent of its GDP on fertility subsidies. For example, mothers with four or more children get a lifetime exemption from income tax.[46] But they have only managed to raise the fertility rate from roughly 1.3 to 1.5.[47] Though this is substantial, it's far from reaching even the replacement rate. Even Hungarian levels of fertility subsidies wouldn't suffice to avert stagnation.

Finally, we could avert stagnation if we develop breakthrough technology in time. We might develop artificial general intelligence (AGI) that could replace human workers—including researchers.[48] This would allow us to increase the number of "people" working on R&D as easily as we currently scale up production of the latest iPhone. If we get to AGI before we stagnate, then longterm stagnation is not an issue; instead, as I argued in Chapter 4, we should then expect technological progress to advance much more rapidly, and we should worry instead about the possibility of value lock-in. Though I think there's a significant chance we will develop AGI this century, we should not be confident that we will do so—AGI might just be very hard.[49]

Advances in biotechnology could provide another pathway to rebooting growth. If scientists with Einstein-level research abilities were cloned and trained from an early age, or if human beings were genetically engineered to have greater research abilities, this could compensate for having fewer people overall and thereby sustain technological progress. But in addition to questions of technological feasibility, there will likely be regulatory prohibitions and strong social norms against the use of this technology—especially against the most radical forms, which would be necessary to multiply effective research efforts manyfold. Human cloning is already within technological reach, but as a global society we've decided not to go forward with it—which may well be for the best, as human cloning could plausibly increase the risk of bad value lock-in.[50]

In sum, if we neither develop and deploy breakthrough technology in time nor see a renewed population boom, it doesn't look like we'll be able to keep quadrupling research effort. In that case, stagnation seems likely.

How Long Would Stagnation Last?

If we entered a period of stagnation, how long would it last? We've seen that rebooting growth might be very hard: there's only so far we can go with policies to reduce scientific bureaucracy and increase the fraction of the population devoted to research, and it's proved difficult for governments to encourage larger families. Might technological stagnation therefore continue indefinitely into the future?

This seems possible but unlikely to me. The key consideration is that getting out of stagnation requires only that one country, at one point in time, is able to reboot sustained technological progress. And if there are a diversity of societies, with evolving cultures and institutional arrangements over time, then it seems likely that one will manage to restart growth.

We've seen this dynamic play out in economic history. In Europe, the Middle Ages was a long period of stagnation. A study of England, where we have the best data, shows that productivity growth, a measure of technological progress, was literally zero from 1250 (when the data start) to 1600.[51] But this stagnation did not last.

Similarly, even if the world enters a period of stagnation, as long as just one society can hit on a sustainably high-growth culture, then the world as a whole will start to technologically advance again. We saw that one major reason for expecting stagnation is that fertility rates are declining, but this could easily change in the future. If some culture particularly values large families and this trait is sustained, that culture would grow to become a progressively larger proportion of the world population over time.

In that case, a single sustained high-fertility culture would ultimately drive global population growth. To see this, suppose that the global population plateaus but a subculture constituting just 0.1 percent of the population continues growing at 2 percent per year. After 350 years, that subculture would amount to more than half of all the people in the world, and the global population growth rate would now be 1 percent per year. After 450 years, the large majority of the population would belong to that subculture, and the global population growth rate would be close to 2 percent per year. If this high-fertility subculture also prizes scientific inquiry, then technological progress may resume.[52]

However, even if stagnation is unlikely to be permanent, there are a number of reasons why it might last for centuries or even millennia. First, as I argued in Chapter 4, to a significant extent we are already living in a single global culture. If that culture develops into one that is not conducive to technological progress, that could make stagnation more persistent. This partly undermines the "diversity of cultures" argument I just gave.

We've already seen the homogenising force of modern secular culture for some high-fertility religious groups. Consider the American Mormons. They're famous for their large families, and until recently, commentators projected that they would grow rapidly as a proportion of the American population.[53] But over time, the Mormon fertility rate has fallen in parallel to the overall American one; now the Mormon fertility rate is just barely above replacement.[54] This seems to be part of a more general, structural pattern. Across many countries, subpopulations, and religious groups, fertility rates have fallen in parallel over the last few decades.[55] While some groups have maintained a higher level of fertility, if the downward trend continues, their fertility rates, too, could fall below replacement, and we would see global population decline.

A single global culture could be especially opposed to science and technology if there were a world government. There would then no longer be competition between countries, so one major motivation behind technological innovation—ensuring greater economic and military power than one's rivals—would be gone. Other motivations to innovate might not be forthcoming because technological change is often disruptive. It can put people out of jobs—think of the Luddites. And it can threaten society's elites: one hypothesis for why the Islamic Golden Age came to an end is that there was a rise in a particular antiscientific religious ideology that helped political elites to entrench their power.[56] Such forces could result in a society opposed to technological innovation.

A second reason why stagnation might last a long time is population decline. As we've seen, global population will plausibly not just plateau but shrink. Fertility rates almost everywhere are falling to substantially below 2. At 1.5 children per woman (roughly the average in Europe), within five hundred years the world population would fall from ten billion to below one hundred million; at one child per woman (roughly the fertility rate in South Korea), the world population would fall to one hundred million within two hundred years.[57]

In this situation, the bar for an outlier culture to restart technological progress is much higher. For example, they'd have to sustain high fertility rates for a long time to get world population back up to ten billion and beyond—a large enough population, with enough researchers, to start driving substantial new technological advances again. That's hard, and a lot can happen in that time.[58] Other one-off gains also become less potent. If a country can implement some policies to make researchers ten times more effective, that still might not suffice to restart growth if the world population has fallen to one hundred million. The deeper you've fallen, the harder it is to get out, and the expected length of stagnation would be greater.

The world population could also decrease dramatically as a result of a global catastrophe, like those discussed in the last two chapters. If a nuclear war or pandemic wiped out 99 percent of the world's population, then, as discussed in the last chapter, we'd likely be able to recover industrial civilisation. But the dramatic population reduction would again make further technological progress very difficult—and the bar for an outlier culture to restart technological progress much higher.

Overall, we don't know just how long stagnation would last. It's possible that stagnation would be short, lasting only a century or two, but it's also possible that it would be very long. Perhaps a stagnant future is characterised by recurrent global catastrophes that repeatedly inhibit escape from stagnation; perhaps cultural norms that are inconducive to progress become globally prevalent and are very persistent; perhaps we end up exhausting all recoverable fossil fuels in a stagnant future and the resulting extreme climate change further impedes growth. If some of these come to pass, then stagnation could potentially last for tens of thousands of years.

Taking this uncertainty fully into account means that the expected length of stagnation could be very great indeed. Even if you think it's 90 percent likely that stagnation would last only a couple of centuries and just 10 percent likely that it would last ten thousand years, then the expected length of stagnation is still over a thousand years.

Stagnation from a Longtermist Perspective

How bad would stagnation that lasts centuries or millennia be? Clearly, during the period of stagnation, people would be much poorer than they could have been if technological progress had continued. Still, one argument you could make is that, as long as growth restarts at some point, then a period of stagnation is not close in importance to extinction or the lock-in of bad values. Just as a growth slowdown might delay us by a decade, a period of stagnation might delay us by a thousand years. But, so the argument goes, whether the delay is ten years or a thousand years, it's pretty minor compared to the millions, billions, or trillions of years ahead of us.[59]

However, what this argument misses is that a centuries-long stagnation could have a major effect on both future values and the probability of civilisation's survival. First, the values that would guide the future after a thousand years of stagnation would probably be very different from the values that are predominant today, simply because there would have been a thousand years of moral change. Would this be a good or a bad thing? There are a number of considerations.

One argument for expecting moral progress during stagnation is that, over time, people generate new moral ideas, make moral arguments, run campaigns, and convince others. And perhaps this process continues whether or not there is technological change. If so, then a thousand-year delay in technological progress would give time for moral progress to continue. The values that would guide the world a thousand years from now would therefore probably be better than the values that guide the world today.

On the other hand, you might expect moral regress if you think the values that guide the world today are unusually good. We've already seen some ways in which this is true compared to history: the global abolition of slavery was unprecedented and, as we've seen, did not seem inevitable. Similarly, there are far more people living in democracies today than at any point in history, and, globally, women now have greater autonomy and political power than ever before. Perhaps over a period of stagnation these moral advances would be lost.

Here are two reasons why this might happen. First, perhaps, as political economist Benjamin Friedman argues, people are more morally motivated in times of economic growth.[60] When the economy is growing, everyone can be better off than they were in the past. This means, Friedman argues, that citizens will worry less about how their life compares to the lives of people around them and will be more supportive of generous, open, and tolerant social policies. And if you look at the historical record, he claims, countries tend to make moral progress—becoming fairer, more open, and more egalitarian—during higher-growth periods, and they tend to morally regress during periods of stagnation.

A second reason ties back to our earlier discussion of cultural evolution. When technological innovation is possible, there are great economic gains to be had from critical thinking and scientific inquiry; and since economically successful cultures gain more members, cultural evolution currently selects for traits conducive to science. As a side effect, so this argument goes, we apply our critical capacities to moral issues, too, and therefore make moral progress. In a stagnant world, the economic reasons to engage in critical thinking and scientific inquiry would be much weaker. Instead, other values would be selected for, such as those favouring hierarchy and conformity, which have guided so many societies in the past.

Even more important than the values during stagnation are whatever values will eventually get the world out of it—for these are the ones that will become predominant in the longer term. These aren't necessarily values that prize critical thinking and inquiry. For example, the prevailing moral worldview could simply be whatever one most champions very high fertility; perhaps this would be a worldview with very inegalitarian gender norms. Or it could be whatever worldview is most willing to break social taboos in the pursuit of economic gains. Perhaps the worldview of whichever country is first willing to use human cloning and genetic engineering will dominate. There's no reason at all to expect this to be an egalitarian and democratic society rather than a fascist or authoritarian regime.

This is all speculative, and I'm not sure which of these perspectives on future moral progress is more correct. I see the questions of whether we should expect values to get better or worse into the future and under what conditions, as crucial and open. At the moment, the issue is extremely underexplored, so I won't draw any strong conclusions.[61]

A different consideration is more clear-cut: a long period of stagnation could substantially increase the probability of extinction or civilisational collapse. As I mentioned in the introduction to this chapter, it matters whether the level of technological advancement is *sustainable*. Had we stayed stuck in 1920s technology, even if we drove our cars less and rode our bikes more, and even if we all stopped eating beef, we still would have inexorably emitted large amounts of carbon dioxide, and we would have eventually burned through all the fossil fuels we could recover. Extreme climate change would have been unavoidable, as would a decline of standards of living as we ran out of carbon to burn.

The only way we got out of that unsustainable state was by inventing ways to produce clean energy. Once we started burning fossil fuels, further technological progress was the only hope for giving us a shot at averting a climate catastrophe without falling back to preindustrial levels of material hardship. And even today, when clean energy is finally available at viable cost, further progress can reduce the cost of decarbonisation and enable us to decarbonise more sectors of the economy. In short, innovation may well be crucial for incentivising countries to adopt the stringent climate change–mitigation policies we need.

A similar consideration applies to the risk of extinction: we may be about to enter an unsustainable state. We are becoming capable of bioengineering pathogens, and in the worst case engineered pandemics could wipe us all out. And over the next century, in which technological progress will likely still continue, there's a good chance we will develop further, extremely potent means of destruction.

If we stagnate and stay stuck at an unsustainable level of technological advancement, we would remain in a risky period. Every year, we'd roll the dice on whether an engineered pandemic or some other cataclysm would occur, causing catastrophe or extinction. Sooner or later, one would. To safeguard civilisation, we need to get beyond this unsustainable state and develop technologies to defend against these risks.

As a result, stagnation could plausibly be one of the biggest sources of risk of extinction or permanent collapse that we face. To illustrate, consider that my colleague Toby Ord puts the risk of human extinction this century from engineered pandemics at around 3 percent.[62] Per-century risk during a period of stagnation might be lower if we adapt with policies like better government regulation of biolabs—or it might be higher if we invent even more destructive technology, or because there is greater potential for conflict in a zero-sum society. But suppose that we got per-century risk down to 1 percent during the period of stagnation and that the period of stagnation lasted for a thousand years. If so, total extinction risk added by stagnation would be around 10 percent; even if stagnation only has a one-in-three chance of occurring, that makes the risk from stagnation comparable in size to the 3 percent risk from engineered pandemics this century.[63]

Earlier, I suggested that civilisation's technological advance is like a climber scaling a sheer cliff face. With a burst of energy, we could press on and reach safety at the summit. But as we've seen, this climber is growing tired, and if they stop entirely, then it might be only a matter of time before they fall.

At this point, I hope I've convinced you that there are real things we can do to predictably affect the very longterm future. We can steer civilisation onto a better trajectory by delaying the point of value lock-in or by improving the values that guide the future. And we can ensure that we have a future at all by reducing the risks of extinction, collapse, and technological stagnation.

In the next part of the book, I tackle two questions that affect how we should prioritise these two ways of affecting the long term. Why should it matter if civilisation's life has been cut short? And is future civilisation, on balance, more good than bad? The answers to these questions determine whether we should focus on trajectory changes or on ensuring survival, or on both. So let's turn to them.

PART IV
ASSESSING THE END OF THE WORLD

CHAPTER 8

Is It Good to Make Happy People?

Derek Parfit

Derek Parfit was one of the most creative and influential moral philosophers of the last century, a machine for turning coffee into philosophical insights.[1] He lived almost all of his life in educational institutions, attending Eton on a scholarship before studying history at Oxford, then winning a prize fellowship at All Souls College. All Souls might be the most exclusive research institute in the world; there are no undergraduates and fewer than ten graduate students at any one time.[2] The qualifying tests for the fellowship have been called "the hardest exam in the world"[3]: twelve hours of domain-specific and general questions and prompts such as, "What is a number?" "Can we be forced to be free?" and even "Defend tweeting." Up until recently, there was a further three-hour exam that simply presented you with a single word, such as "water," "novelty," or "reproduction," and required you to write a full essay on the topic.[4] After receiving the fellowship at age twenty-four, Parfit spent the next forty-three years at All Souls and never completed any of his philosophy degrees.

He was utterly single-minded in his pursuit of improving our moral understanding. In the latter half of his life, he would take every opportunity to save time on anything that wasn't philosophy: literally running between seminars, wearing the same outfit every day (black trousers and a white shirt), and eating the same easy-to-prepare vegetarian meals (cereal with yogurt and blackberries for breakfast; for dinner, raw carrots, romaine lettuce, celery dipped in peanut butter or hummus, followed by tangerines and apples). He would read philosophy while brushing his teeth. The coffee he drank was instant, filled from the hot water tap so that he didn't have to wait for the kettle to boil. As *New Yorker* journalist Larissa MacFarquhar noted in her profile of him, "The driving force behind Parfit's moral concern was suffering. He couldn't bear to see someone suffer—even thinking about suffering in the abstract could make him cry."[5]

His capacity for philosophy and his generosity were boundless. As a graduate student, I once provided him with comments on a draft article of his. I thought these were rather lengthy at three thousand words; even so, a typical response from a senior professor would be "Thanks." Parfit, however, quickly responded with nine thousand words, about the length of a typical journal article. He apologised for the length, telling me he had taken some time to shorten it. Tragically, he passed away in early 2017.

Parfit inaugurated several new areas of moral philosophy. The one that has most shaped my worldview, and which is covered in this chapter, is *population ethics*—the evaluation of actions that might change who is born, how many people are born, and what their quality of life will be. Secular discussion of this topic is strikingly scarce: despite thousands of years of ethical thought, the issue was only discussed briefly by the early utilitarians and their critics in the late eighteenth and nineteenth centuries, and it received sporadic attention in the years that followed.[6] The watershed moment came in 1984 with the publication of Parfit's book *Reasons and Persons.*

Population ethics is crucial for longtermism because it greatly affects how we should evaluate the end of civilisation. Parfit himself recognised this, writing, at the very end of *Reasons and Persons*,

> I believe that if we destroy mankind, as we now could, this outcome would be much worse than most people think. Compare three outcomes:
>
> (1) Peace.
>
> (2) A nuclear war that kills 99% of the world's existing population.
>
> (3) A nuclear war that kills 100%.
>
> Outcome (2) would be worse than (1), and (3) would be worse than (2). Which is the greater of these two differences? Most people believe that the greater difference is between (1) and (2). I believe that the difference between (2) and (3) is very much greater.[7]

The reason that Parfit regarded extinction as far worse even than a catastrophe that killed 99 percent of the global population is that extinction would not just involve the deaths of the eight billion people alive today; it would also prevent the existence of all the people who otherwise would have lived in the generations to come. The end of civilisation would mean the absence of trillions upon trillions of people who would otherwise have been born. Parfit concluded that preventing the existence of a happy and flourishing life is a moral loss; the loss from human extinction is therefore vast. In later work, he concluded that "what now matters most is that we avoid ending human history."[8]

When I first came across the idea of regarding the prevented existence of a happy life as a moral loss, I found it bizarrely unintuitive. Over time, the force of arguments in favour of this view changed my mind. Indeed, this is one of the most significant ways in which moral philosophy has changed my ethical views, and I think that Parfit's arguments, and the arguments of others in the field of population ethics, are among the most important contributions of moral philosophy of the last century.

In this chapter, I'm going to explain these arguments and defend Parfit's view that, provided a person had a sufficiently good life, the world would be a better place in virtue of that person being born and living that life. Crucially, this isn't the claim that an additional person might make the world better by enriching the lives of others; instead, it's the claim that having one extra person in the world is good in and of itself, if that person is sufficiently happy. So, throughout most of this chapter, I will bracket questions around the harms that people might impose by using resources or producing pollution, or the benefits they might produce by creating life-saving inventions. While these are important factors, I am concerned not with the instrumental effects of additional people but with the question of whether adding sufficiently happy people is noninstrumentally or *intrinsically* good. I also do not claim that we are morally required to bring more happy people into existence, or that we're blameworthy if we fail to do so—just that, all other things being equal, having more happy people makes the world a better place.

Before we begin, allow me a few caveats. The first is that this is going to be the most theoretical chapter in the book. Population ethics is recognised as one of the most complex areas of moral philosophy, and at universities it is normally studied only at the graduate level. To my knowledge, these ideas haven't been presented to a general audience before. But they are of such great importance for thinking about the longterm future that I simply must discuss them. I will do my best to simplify things, but the subject matter itself is often complex and confusing. As will become clear in what follows, all theories of population ethics have some unintuitive or unappealing implications. The task is to decide which unappealing implications we must accept.

Second, I'll talk a lot about people's wellbeing or happiness—I use the terms interchangeably. By this I mean how well or poorly someone's entire life goes, not just how well-off someone is at a specific moment in time. I'll sometimes use numbers to represent how well-off someone is; when I do, I'll use "100" to refer to an extraordinarily good life, happy and flourishing; I'll use "–100" to refer to an extraordinarily bad life, full of misery and suffering; and I'll use "0" to refer to a life that is neither good nor bad from the perspective of the person living it. Crucially, I'm not assuming anything about the nature of wellbeing. A good life could consist of joyful experiences, or meaningful accomplishments, or the pursuit of knowledge and beauty, or the satisfaction of one's preferences, or all of these things combined. Whichever of these views we have, we need to think about population ethics.

Third, in this chapter I'll talk about lives that are below neutral wellbeing—lives such that it would be better, for the people living them, if they had never been born. This can be a disturbing idea, and I've met people who claimed that it is simply not possible for a life to be below neutral wellbeing. But that cannot be correct. Recall the most extreme suffering you have ever experienced and imagine a life that consisted of nothing but that suffering. Would you choose to live that life if the alternative was nonexistence? If you answer no, that suggests you agree that, in principle, a life can be below neutral wellbeing.

Importantly, that someone has a life with below-neutral wellbeing does not entail that their life is not worth living. Even if a person is persistently depressed, they can make a great contribution to the world by being a good friend or family member, by being a doctor or a scientist producing lifesaving research. And if someone has below-neutral wellbeing at a particular time, that does not mean that their whole life is below neutral. Almost everyone goes through periods of sadness and depression, but that does not mean that their whole life has been negative for them.

Fourth, when I talk about populations, I mean total populations: not just how many people are alive at a specific time but all people across all time.

Finally, to test different theories in population ethics, I will evaluate what they say about how we should compare different populations. In practice, we will probably never get to make choices between such populations, but considering these hypothetical cases is still the best way to assess whether a theory is true. As I hope will become clear by the end of the chapter, this is not all merely idle philosophical speculation: it really does matter, for ordinary people and governments, which theory of population ethics is true.

With these clarifications established, we can look at some different perspectives on population ethics.

The Intuition of Neutrality

The view that the world is made better by having more people with sufficiently good lives is often regarded as unintuitive. Philosopher Jan Narveson put it in slogan form: "We are in favour of making people happy, but neutral about making happy people."[9] One of my PhD supervisors, economist-turned-philosopher John Broome, called this the "intuition of neutrality"—the idea that bringing someone with a good life into existence is a neutral matter.[10] While writing a book on population ethics, Broome struggled for over a decade trying to justify it before grudgingly accepting that it had to be abandoned.[11] I, too, had this intuition and was reluctant to reject it.

You might feel this intuition if you reflect on how you'd reason when deciding whether to have a child. You might think through many reasons in favour or against: whether it would make your life and the lives of your family members happier and more meaningful; whether the child would, through their good deeds, go on to improve society. Perhaps you would think about your child's carbon footprint. But you might think it would be odd to claim that the fact that the child would have a good life is itself a reason to have a child.

If you endorse the intuition of neutrality, then ensuring that our future is good, while civilisation persists, might seem much more important than ensuring our future is long. You would still think that safeguarding civilisation is good because doing so reduces the risk of death for those alive today, and you might still put great weight on the loss of future artistic and scientific accomplishments that the end of civilisation would entail. But you wouldn't regard the absence of future generations in itself as a moral loss.

However, there are many situations where the intuition of neutrality is very unintuitive. This is clearest when we imagine lives consisting entirely of misery and suffering. Imagine a life that, from birth till death, consists only of agony and anguish; imagine, for example, someone who continually felt like they were being burned alive. And imagine that you know you could have a child who would live such a life. It seems entirely obvious to me that having this child would be a bad thing to do.

For this reason, most philosophers who endorse the intuition of neutrality endorse an asymmetry. They believe that, although it's not good to bring a new person with a happy life into existence, it is bad to bring a new person with an unhappy life into existence. But it's not clear how we can justify this asymmetry, though many philosophers have tried. If we think it's bad to bring into existence a life of suffering, why should we not think that it's good to bring into existence a flourishing life? I think any argument for the first claim would also be a good argument for the second.

This idea becomes more plausible when we think of lives that are sufficiently good. For example, I have one nephew and two nieces, who are all still young. They are happy children, and if I imagine this happiness continuing into their futures—if I imagine they each live a rewarding life, full of love and accomplishment—and ask myself, "Is the world at least a little better because of their existence, even ignoring their effects on others?" it becomes quite intuitive to me that the answer is yes. If so, the intuition of neutrality is wrong.[12]

Philosophers often claim that the intuition of neutrality is part of the "commonsense" moral view, but really, it's not clear that this is true. The only psychological study on this topic asked participants how much better or worse the world would be if one new person were added to it.[13] In one variant of the question, it was stipulated that the new person "would be extremely happy and live a life full of bliss and joy"; in the other variant, the new person "would be extremely unhappy and live a life full of suffering and misery." It was emphasized that there would be no other negative or positive impacts on others from the existence of this person.

The authors of the study found that people, on average, think that it's a good thing to bring a new happy person into existence and that it's a bad thing to bring a new unhappy person into existence. Moreover, these judgments were symmetrical: the experimental subjects were just as positive about the idea of bringing into existence a new happy person as they were negative about the idea of bringing into existence a new unhappy person. That is, those surveyed did not have the intuition of neutrality.

Clumsy Gods: The Fragility of Identity

A second argument against the intuition of neutrality again comes from Parfit.[14] He noted that our existence in the world is exceptionally unlikely, and the identity of future people is exceptionally fragile, and that major ethical implications follow from this.

Time travel stories often illustrate how the present can be highly dependent on small decisions in the past. In *Back to the Future*, for example, Marty McFly goes back in time, takes his mother to a high school dance, sets her up with his father, and helps his father defeat Biff, the school bully. Though his parents ultimately marry, preserving Marty's existence, when he returns to the present there are some major changes to his life: his dad is a successful writer, and Biff—instead of bullying his father, as he did before Marty time travelled—cleans his family's car. But I think that if we consider the changes to the past that Marty McFly made, the changes to his present would have been *much* greater than the film suggests.

Consider that a typical ejaculation contains around two hundred million sperm. If any of the other two hundred million sperm had fertilised the egg that you developed from, then you would not have been born. Instead someone else—with 75 percent of your genes—would have been born in your place. A one-in-two-hundred-million event involves extreme luck. So, as much as I'm sure you don't want to think about such things, if your father's ejaculation had occurred just milliseconds earlier or later, it would almost certainly have been a different sperm that fertilised your mother's egg. And so any event that affected the schedules of your biological mother and father on the day that you were conceived, even if only by a tiny amount—such as a longer line at the supermarket or an additional car ahead of them on their way home from work—would have prevented you from being born.[15] When Marty McFly returns to the present, his siblings are the same people they were before his time travel adventure (if more successful). But if he really had gone back in time and made any changes at all to his parents' lives, he would have changed his siblings' identities—and, paradoxically, his own!

If someone else had been born in your place, this would have had countless knock-on events. Your sibling's time of birth would probably have been different, as would their personality. They would have altered how your parents, and people who interacted with your parents, behaved over the course of decades. And all those interactions would have altered the timings of countless other reproductive events, changing which sperm met the eggs and altering the identities of the babies who were subsequently born. These changes would also have impacted the timing of further reproductive events, until at some point in the future, the identities of everyone who is born is different than they would have been. And this is all because of small decisions like which route home your parents took from work one day. I dedicated my first book, *Doing Good Better*, to Peter Singer, Toby Ord, and Stanislav Petrov, and I said that "without [them] this book would not have been written." But the book also would not have been written were it not for Jesus, Hitler, or any random English peasant in the fifteenth century.

In time travel stories, small actions in the past often result in radical changes in the present. But we rarely think about the fact that small actions today can have dramatic effects on the future.[16] Do the very longterm consequences of our actions fade out over time, like ripples on a pond? No. Rather, every year, like clumsy gods, we radically change the course of history. For example, if you live in a city, then by choosing to take public transport to work and back, rather than drive, over the course of a year you will ever-so-slightly impact the schedules of tens of thousands of people over hundreds of days. Statistically, it's likely that, on one out of those tens of thousands of person-days, the person you impacted had sex and conceived a child later in that day,[17] and you affected, ever so slightly, the timing of that conception, changing which sperm met the egg and thus changing who was born. That different person will then impact the schedules of millions of other people, changing what children *they* have, and so on, in an identity cascade. Past a certain date, everyone who is ever born will be different from who *would* have been born if you had chosen to drive instead, and the entire course of future history will be different. Wars will be fought that would never have been fought; monuments built that would never have been built; works of literature written that would never have been written. All because you chose to take the bus rather than drive.

The fragility of identity has important philosophical implications. Suppose that the world's governments decide to end fossil fuel subsidies. Intuitively, we might think that by reducing climate change, this decision improves the lives of specific people in the future who would exist either way. But this is incorrect. A large policy change like this would impact everyone in the world: it would make petrol more expensive and so would affect traffic globally. It would change everyone's schedules and, by affecting the timing of conceptions, within a few years it would change the identities of almost every person who is born. From a few years onwards, the new population will be made up of entirely different people than those who would have otherwise existed.

These people will be better off than those who would have existed had we kept fossil fuel subsidies, but they will be *different* people. And according to the intuition of neutrality, we cannot make the world better by adding new people. So we cannot say that ending fossil fuels is good because it benefits future generations.

Consider two people, Alice and Bob. If we keep fossil fuel subsidies, Alice will be born in 2070. If we end fossil fuel subsidies, Alice will not be born and Bob will be born instead. Both have happy lives, but, because climate change will be less extreme without fossil fuel subsidies, Bob will be happier than Alice would have been. According to the intuition of neutrality, we do not have reason to ensure that Bob exists rather than Alice. According to the intuition of neutrality, preventing Alice's existence is neither good nor bad, and bringing Bob into existence is also neither good nor bad. So doing both at once is neither good nor bad.

This implication of the intuition of neutrality seems wrong. Intuitively, the fact that ending fossil fuel subsidies will change the identities of future people just doesn't matter, morally. The reasons the world's governments have to end fossil fuels are just as strong whether or not they will change who exists in the future. Ending fossil fuel subsidies makes the future better. But it does so by *creating* a population that is *made up of completely different people* than the population that would have existed otherwise. Adding new people cannot, then, be a neutral matter.

Why the Intuition of Neutrality Is Wrong

So far, we've seen arguments for thinking that the intuition of neutrality is much less intuitive than it might first seem. But there is also a powerful argument in favour of that view based on what is a surprisingly simple piece of logic.[18]

Suppose that a couple are deciding whether or not to have a child. Because of a vitamin deficiency that the mother is currently suffering from, the child they conceive will certainly suffer from migraine: every few months, for their entire life, they will suffer a debilitating headache and have fatigue and brain fog for several days afterwards. But other than this, the child will live a good and full life. According to the intuition of neutrality, it is a neutral matter whether or not these parents have this child: the world is equally good either way.

Now suppose that the parents also have the option of having the child a few months later. At that later point, the mother will no longer suffer from the vitamin deficiency, and the child they conceive will not suffer from migraine as a result. Let's call the option of having no child "No Child";

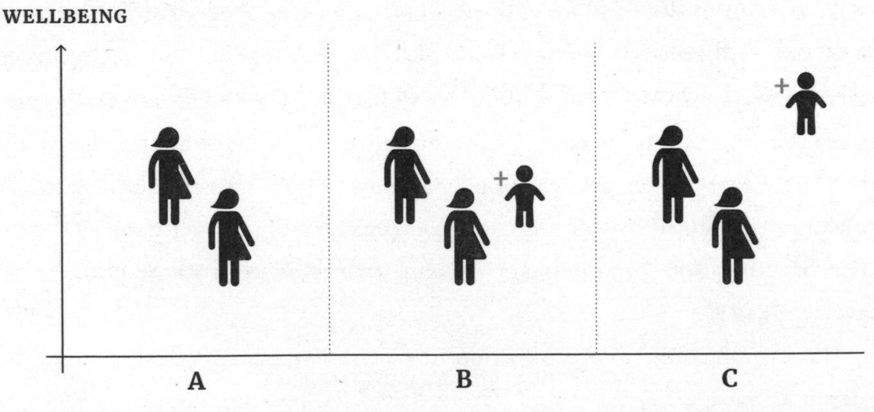

Figure 8.1. Consider a choice between options A, B, and C. A is the option of having no child, B is the option of having a child with migraine, and C is the option of having a migraine-free child. This choice poses a problem for the intuition of neutrality.

"Migraine" is the option of having a child with migraines; and "Migraine-Free" is the option of having a child without migraine (Figure 8.1).

It seems obvious that, as long as there are no other considerations in play, if the parents have the choice, they should choose to have a child that is migraine-free over a child with migraine. That is, Migraine-Free is better than Migraine. But if so, then the intuition of neutrality must be wrong: having a child cannot be a neutral matter.

To see this, first compare No Child and Migraine. According to the intuition of neutrality, the world is equally good either way, whether the parents decide to have no child or to have the child with migraine. That is, No Child is equally good as Migraine.

Second, compare No Child with Migraine-Free. According to the intuition of neutrality, the world is equally good either way, whether the parents decide to have no child or to have the child without migraines. That is, No Child is equally good as Migraine-Free.

However, if No Child is equally good as Migraine, and No Child is equally good as Migraine-Free, then Migraine and Migraine-Free must be equally good. But we know that having the child with migraine is worse than having the child without migraine: the two outcomes are exactly the same except that, in one outcome, one person has more suffering in their life. The intuition of neutrality has led us into a contradiction.

Various philosophers have now spent several decades playing argumentative whack-a-mole trying to avoid the problems with the intuition of neutrality.[19] It's impossible to do justice to all these potential responses, especially as the ensuing discussion gets very technical very quickly. But, in my view, all proposed defences of the intuition of neutrality suffer from devastating objections.

If we give up on the intuition of neutrality, what should we have instead? Parfit himself didn't know. He called the quest for the correct theory of population ethics the quest for "Theory X."[20] Let's turn to a few candidates for such a theory.

The Average View

You might be tempted to suggest that what's important is to try to increase a population's *average* wellbeing. In this view, it's better to have fifty thousand people at +60 happiness than to have four hundred thousand at +40 happiness. This is a view that is often assumed, implicitly or explicitly, by economists, and surveys suggest that it seems to have a basis in common sense.[21]

However, though philosophers agree on very few things, one of the things they do agree on is that the average view is wrong. It suffers from an absolute litany of problems. Here are just two. First, if the world consisted of a million people whose lives were filled with excruciating suffering, one could make the world better by adding another million people whose lives were also filled with excruciating suffering, as long as the suffering of the new people was ever-so-slightly less bad than the suffering of the original people. (This is a thought experiment that Parfit presented and referred to as "Hell Three.") If the original one million people have –100 wellbeing, then in the average view, adding a further million people at wellbeing level –99.9 is a good thing because it brings up the average. But this is absurd.

We can illustrate this using a box diagram (see Figure 8.2), which is a way to compare different populations. The boxes represent populations. The width of each block shows the number of people in the corresponding population over all time; the height shows their lifetime wellbeing. Lives above the horizontal line have positive wellbeing; those below have negative wellbeing.

The second problem is that in the average view it can be better to create new lives filled with suffering than to create new very happy lives. Suppose

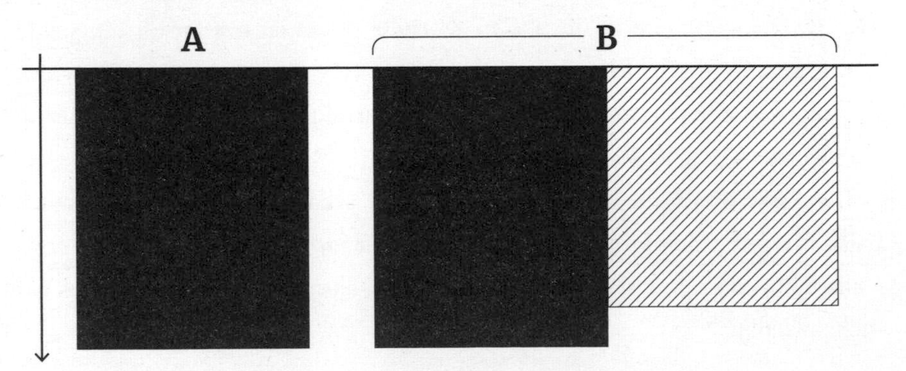

Figure 8.2. Box diagram illustrating the Hell Three *argument against the average view. Populations A and B both consist only of people with such horrible lives that they would prefer to never have been born. Population B differs from A only in that it contains a larger number of terribly suffering people. The average view says that B is* better *than A because it has higher average wellbeing.*

that the world consists of ten billion people at wellbeing 100. We could either add ten million people in excruciating suffering, at wellbeing –100, or three hundred million people with happy and flourishing lives at wellbeing 90. Adding the three hundred million people at wellbeing 90 would bring down the average by more than adding the ten million people at wellbeing –100. So, in the average view, it would be better to add the ten million lives of excruciating suffering.[22] This, again, is absurd. Given these problems, we should not be tempted to endorse the average view (see Figure 8.3).

The Total View

If we reject both the intuition of neutrality and the average view, the most natural alternative is the *total view*. In this view, one population is better than another if it contains more total wellbeing.

The basic motivation for the total view is simply that *more of a good thing is better*.[23] Good lives are good. More of a good thing is better. So increasing the number of good lives makes the world better.

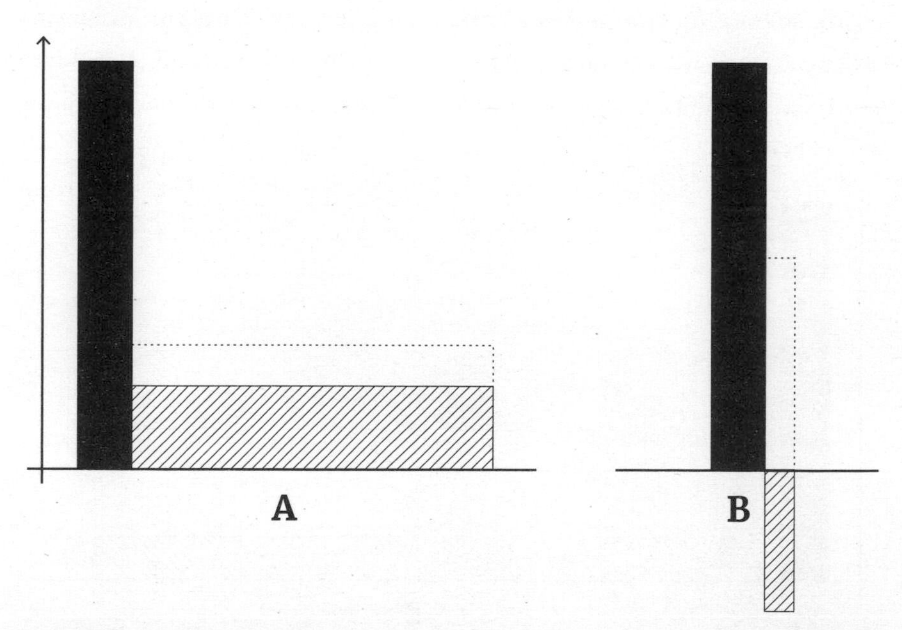

Figure 8.3. Box diagram illustrating that, on the average view, creating lives with negative wellbeing can be better than creating lives with positive wellbeing. Starting from the happy population represented by the black bar, adding a large number of people with lower, but still positive, wellbeing (resulting in population A) reduces average wellbeing by more than adding a sufficiently smaller number of people with lives that are so horrible it would have been better for them to never have been born (resulting in population B).

The primary objection to the total view is as follows. Consider two worlds: we'll call the first Big and Flourishing and the second Enormous and Drab. Big and Flourishing contains ten billion people, all at an extremely high level of wellbeing. Enormous and Drab has an extraordinarily large number of people, and everyone has lives that have only slightly positive wellbeing. If the total view is correct then, as long as the number of people in the second world is large enough, we must conclude that the second world is better than the first. The wellbeing from enough lives that have slightly positive wellbeing can add up to more than the wellbeing of ten billion people that are extremely well-off.

Parfit himself thought that this was a deeply unpalatable result, so unpalatable that he called it the Repugnant Conclusion, and the name stuck (see Figure 8.4).[24] Initially, he described those slightly-positive-wellbeing lives as consisting of "listening to Muzak and eating potatoes."[25] Later in his life, his favoured formulation was to imagine these lives as lizards basking in the sun.[26]

The Repugnant Conclusion is certainly unintuitive. Does this mean that we should automatically reject the total view? I don't think so. Indeed, in what was an unusual move in philosophy, a public statement was recently

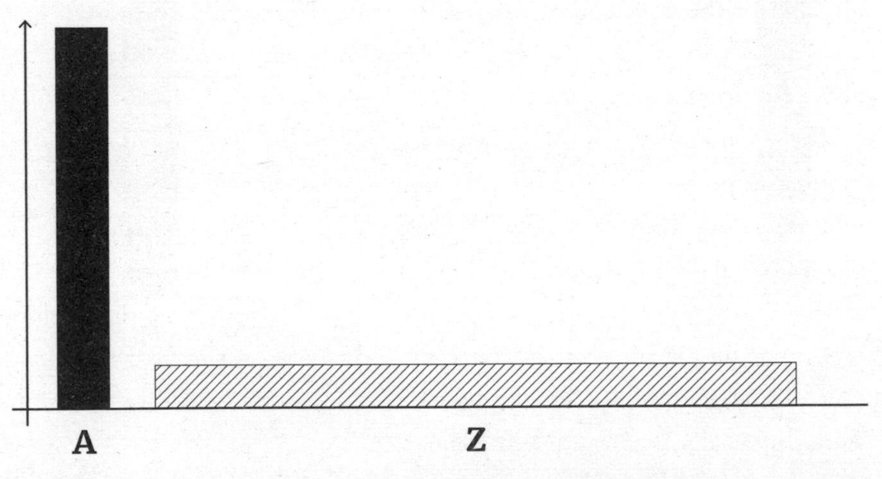

Figure 8.4. Box diagram illustrating the Repugnant Conclusion*: for any happy population (e.g., population A)—no matter how good their lives are—there is a population in which everyone is much worse off (but still enjoys positive wellbeing) but which according to the total view is better because it consists of enough people (e.g., population Z).*

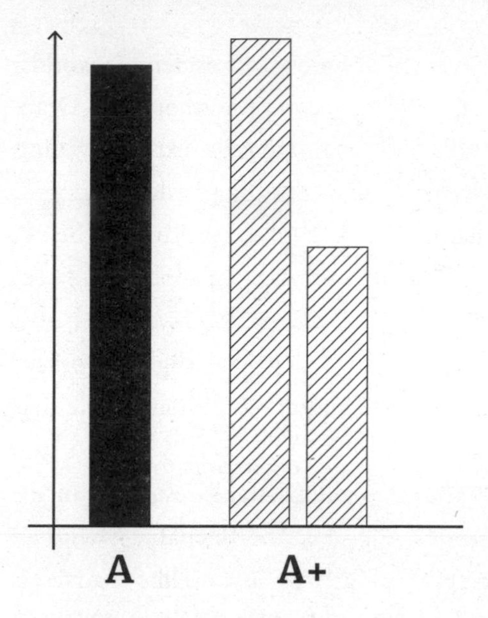

Figure 8.5. According to Dominance Addition, *population A+ is better than population A.*

published, cosigned by twenty-nine philosophers, stating that the fact that a theory of population ethics entails the Repugnant Conclusion shouldn't be a decisive reason to reject that theory.[27] I was one of the cosignatories.

Though the Repugnant Conclusion is unintuitive, it turns out that it follows from three other premises that I would regard as close to indisputable. The first premise is that, if you make everyone in a given population better off while at the same time adding to the world people with positive wellbeing, then you have made the world better. This premise is known as Dominance Addition (see Figure 8.5).[28]

The second premise is that, if we compare two populations with the same number of people, and the second population has both greater average and total wellbeing, and that wellbeing is perfectly equally distributed, then that second population is better than the first. This premise is known (catchily!) as Non-Anti-Egalitarianism (Figure 8.6). The basic idea behind this premise is that equality is not *actively bad*. While some people deny that equality is intrinsically good, to my knowledge no one thinks that equality makes the world *worse*, all other things being equal.

The third premise is that, if one world is better than a second world, which itself is better than a third, then the first world is better than the third. If A > B and B > C, then A > C. This is called Transitivity.

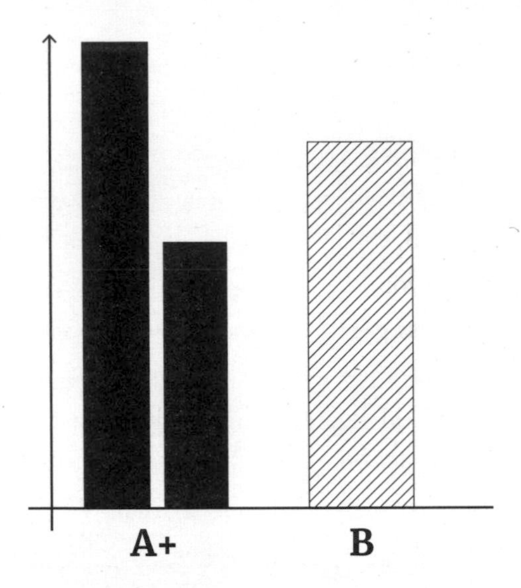

Figure 8.6. According to Non-Anti-Egalitarianism, *population B is better than population A+.*

If we endorse these three premises, then we must endorse the Repugnant Conclusion. To see this, let's combine the two previous diagrams (Figure 8.7).

Consider, first, what I'll call World A: a world of ten billion people who all live wonderful lives of absolute bliss and flourishing. We would, of course, regard this as a very good world. Next, consider World A+. This world differs from A in only two ways. The ten billion people in A+ have even better lives than those in A, and the total population is larger: in A+ there are an additional ten billion people who have pretty good lives, though much less good than the other ten billion people's. So in A+ there are twenty billion people in total.

A+ is better than A for the people who would exist in either world. And the additional ten billion people who would live in A+ have good lives. So

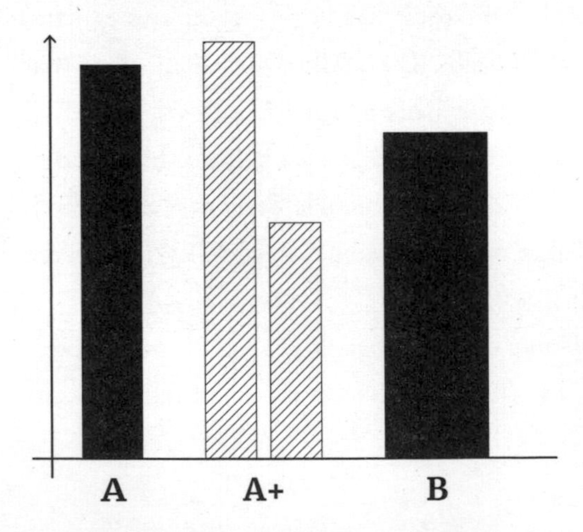

Figure 8.7. Dominance Addition *and* Non-Anti-Egalitarianism *imply that population B is better than population A, assuming that "better than" is a transitive relation.*

we should think that A+ is a better world than A. That's the Dominance Addition premise in play.

Next, consider World B. In this world, there are the same number of people as in A+. But there is no longer any inequality; everyone has the same level of wellbeing. What's more, in World B, the average and total wellbeing are greater than those of World A+. Everyone has equally good lives, and those lives are very good, just a little bit less good than the lives of the residents of A.

On average and in total, people in World B are far better off than the people are in A+, and the distribution of wellbeing is perfectly equal (unlike the very unequal A+). So we should think that World B is better than World A+. That's the premise of Non-Anti-Egalitarianism in play.

Finally, because we thought that B was better than A+, and that A+ was better than A, we should conclude that B is better than A. That's the premise of Transitivity coming in. And if we conclude that B is better than A, then we're concluding that a larger population with a lower average wellbeing is better than a smaller population with greater average wellbeing.

But now notice that we can repeat the process that we just ran through (Figure 8.8).

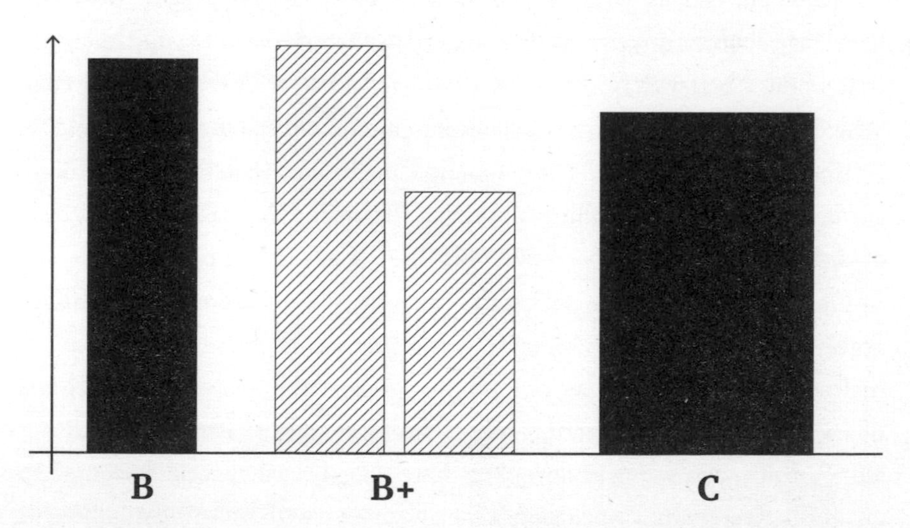

Figure 8.8. As in Figures 8.5 to 8.7, Dominance Addition *and* Non-Anti-Egalitarianism *imply that population C is better than population B, assuming that "better than" is a transitive relation.*

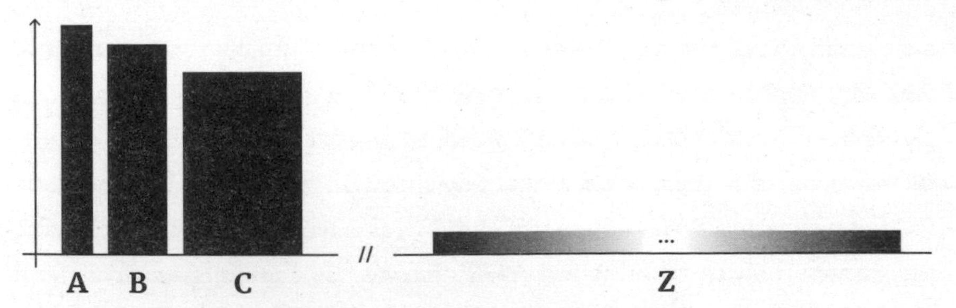

Figure 8.9. As in Figures 8.7 and 8.8, starting from any happy population A, we can construct a series of progressively larger populations B, C, etc. with progressively lower wellbeing, each "better" than the last. We eventually arrive at a very large population Z of barely positive wellbeing, which is better than the original population A. In other words, the seemingly uncontroversial premises Dominance Addition *(Figure 8.5) and* Non-Anti-Egalitarianism *(Figure 8.6), if we also assume that the "better than" relation is transitive, imply the* Repugnant Conclusion *from Figure 8.4.*

We could consider World B+, which makes the people in World B a little bit better off and adds an extra twenty billion people with lives that are pretty good but not as good as the original twenty billion lives. And then we could consider World C, which is just like B+ except that everyone in B+ is equally well-off, at a level of wellbeing that is just a little bit below the best-off people in B+. And so on: we could keep iterating this process over and over, making people's average wellbeing a little bit lower in exchange for making the population larger (Figure 8.9).

We would end up with an enormous number of people with lives that have only slightly positive wellbeing, and we would have to conclude that that world is better than the world we started with, with ten billion lives of bliss. That is, we have arrived at the Repugnant Conclusion.

If you want to reject the Repugnant Conclusion, therefore, then you've got to reject one of the premises that this argument was based on. But each of these premises seem incontrovertible. We are left with a paradox.

One option is simply to accept the Repugnant Conclusion—and perhaps argue that it is not quite as repugnant as it first seems. This is the view that I incline towards. Many other philosophers believe that we should reject one of the other premises instead. Indeed, this was true of Parfit. He was not alone in this, and many philosophers have constructed theories designed to avoid the Repugnant Conclusion. One alternative with prominent adherents is the critical level view.

The Critical Level View

In the critical level view, it's a good thing to bring into existence a good life, but only if that life is *sufficiently good*, above a certain "critical level" of wellbeing.[29] To this, the critical level view adds the idea that it's bad to bring into existence a life that has positive wellbeing but is not very good. This is in contrast to the total view, in which it's always a good thing to bring into existence a life with positive wellbeing.

In the critical level view, adding lives that have low but positive wellbeing is a bad thing.[30] So the critical level view denies the Dominance Addition premise. This view escapes the Repugnant Conclusion (Figure 8.10).

However, the critical level view has its own counterintuitive implications.[31] For example, like the average view, it leads to what's called the Sadistic Conclusion: that it can be better to add to the world lives full of suffering than it is to add good lives (Figure 8.11).

To see this, suppose that 10 represents the critical level of wellbeing. On the critical level view, adding a hundred people at wellbeing level 5 to the population is worse than adding ten people at –30 wellbeing. The critical level view regards the addition of lives that only just have positive wellbeing as a bad thing; so adding enough such lives can result in worse overall wellbeing than adding a smaller number of lives that are full of suffering. This

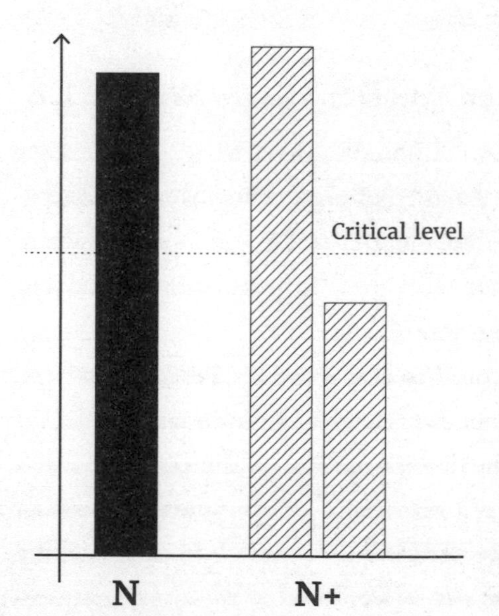

Figure 8.10. Box diagram illustrating that critical level views do not satisfy Dominance Addition. *In the critical level view, adding people whose wellbeing is positive but below the critical level, such as the right bar in population N+, makes the world worse. The left bar in N+ having higher wellbeing than N does not compensate for this negative effect. Therefore, overall, population N+ is worse than population N, contrary to* Dominance Addition.

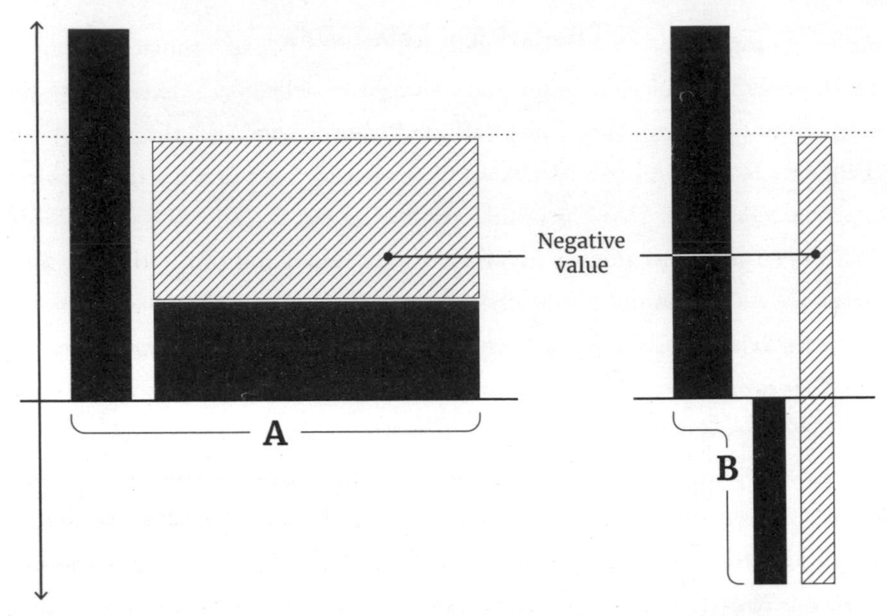

Figure 8.11. Box diagram illustrating that critical level views imply the Sadistic Conclusion. *Consider any population with a level of wellbeing that is positive but below the critical level, such as the right bar in population A. Rather than adding this population of happy people, in the critical level view it is better to add a population consisting only of negative-wellbeing lives, provided that population is sufficiently small—such as the right bar in population B. This is true no matter which population one starts with (such as the left bars in A and B, respectively).*

seems wrong. Like all views in population ethics, the critical level view has some very unappealing downsides.

What You Ought to Do When You Don't Know What to Do

There is still deep disagreement within philosophy about what the right view of population ethics is. I think the balance of arguments favours the total view, but, given how difficult the subject matter is, I'm not at all certain of this. Indeed, I don't think that there's any view in population ethics that anyone should be extremely confident in.

Despite this uncertainty, we still need to act. So we need to know how to act despite our uncertainty. In Chapter 2, I introduced the idea that expected value is the right way of evaluating options in the face of uncertainty. In that chapter I was talking about empirical uncertainty—uncertainty about what will happen. In this context, what we need is an account of decision-making when there's uncertainty about what's of value. In other work,[32] I've argued that, at least in many circumstances, we can extend expected value theory to incorporate uncertainty about value, too. In the case of population ethics, what we should do is figure out what degree of belief we ought to have in each of the different views of population ethics and then take the action that is the best compromise between those views—the action with the highest expected value.

To illustrate this, suppose you assign some probability to both the total view and the critical level view, and for simplicity's sake, let's put all other views to the side. If you're maximising expected value, you end up following a critical level view, though with a lower level for a "sufficiently good" life than if you were certain about the critical level view. Suppose, for example, that you're split fifty-fifty between the total view and the critical level view, with the level for a sufficiently good life set at a wellbeing of 10. Then, if you're maximising expected value, the effective critical level would be halfway between the level of the total view (that is, 0) and the level of the critical level view (that is, 10). Under this moral uncertainty, it's good to bring someone into existence if their wellbeing is greater than 5, but it's not good to bring someone into existence if their wellbeing is less than 5.

My colleagues Toby Ord and Hilary Greaves have found that this approach to reasoning under moral uncertainty can be extended to a range of theories of population ethics, including those that try to capture the intuition of neutrality. When you are uncertain about all of these theories, you still end up with a low but positive critical level.[33]

The Benefits of Having Kids

In rich countries, people generally want to have more kids than they end up having: Americans, for example, want to have 2.6 children on average but have only 1.8.[34] In significant part this is because work and other commitments get in the way. But increasingly, people are starting to see the choice to have children as an unethical one because having children means greater carbon dioxide emissions and faster climate change.[35]

I think this is a mistake. Children have positive effects as well as negative ones. In addition to the direct positive impacts on their family and the friends they will make, when children grow up they contribute to public goods through their taxes, they build infrastructure, and they develop and champion new ideas about how to live and how to structure society. In the last chapter we saw that the recent decline in fertility might lead to a long period of stagnation, extending the time of perils. Having kids can help mitigate this risk.

So far, the knock-on effects of a growing population have clearly been positive, for human beings at least. If they were not, then we would expect the recent dramatic increase in population size to be associated with ever-expanding human misery, but in fact we've seen the opposite. Think about how much worse the world would be if Benjamin Lay, Frederick Douglass, and Harriet Tubman had never existed, or if Marie Curie, Ada Lovelace, or Isaac Newton had never been born. Remember, you are population too![36] If you think you have made the world a better place, then you must think that new people can as well.

In addition to the positive knock-on effects of having children, if your children have lives that are sufficiently good, then your decision to have them is *good for them*. With a sufficiently good upbringing, having a chance to experience this world is a benefit. And, by the same token, if you have grandchildren, you benefit them, too.

Of course, whether to have children is a deeply personal choice. I don't think that we should scold those who choose not to, and I certainly don't think that the government should restrict people's reproductive rights by, for example, limiting access to contraception or banning abortion.

But given the benefits of having children and raising them well, I do think that we could start to once again see having kids as a way of positively contributing to the world. Just as you can live a good life by being helpful to those around you, donating to charity, or working in a socially valuable career, I think you can live a good life by raising a family and being a loving parent.

Bigger Is Better

Population ethics might change how we view the benefits of having a family, but that is not its main implication. The most important upshot of population ethics concerns the question, "How bad is the end of civilisation?" Should we care about the loss of those future people who will never be born if humanity goes extinct in the next few centuries? We now have our tentative answer: yes, it *is* a loss if future people are prevented from coming into existence—as long as their lives would be good enough. So the early extinction of the human race would be a truly enormous tragedy.

In fact, the conclusion that follows is more general than this. If future civilisation will be good enough, then we should not merely try to avoid near-term extinction. We should also hope that future civilisation will be *big*. If future people will be sufficiently well-off, then a civilisation that is twice as long or twice as large is twice as good.

The practical upshot of this is a moral case for space settlement. Though Earth-based civilisation could last for hundreds of millions of years, the stars will still be shining in trillions of years' time, and a civilisation that is spread out across many solar systems could last at least this long. And civilisation could be expansive as well as long. Our sun is just one of one hundred billion stars in the Milky Way; the Milky Way is one of just twenty billion galaxies in the affectable universe.[37] The future of civilisation could be literally astronomical in scale, and if we will achieve a thriving, flourishing society, then it would be of enormous importance to make it so.

That doesn't mean we should pursue space settlement *now*. Space settlement might well be a point of lock-in: the norms, laws, and distribution of power that are present at the time of the first settlers could determine who has access to which celestial bodies and how they are used.[38] By not rushing headlong into space settlement, we preserve option value, ensuring we have time to design systems of governance that don't merely replicate today's injustices far into the future.

And there are more urgent priorities, too. Contemporary efforts to explore the solar system, like the *Curiosity*, *Perseverance*, and *Zhurong* rovers on Mars, can be important for advancing science and for inspiring humanity. But the key practical implication of this chapter is that we should focus on preventing the threats of catastrophe that face us this century, so that we have any chance at all of building a flourishing interstellar society in the centuries that follow.

Moreover, the "if" that all this discussion is based on is a big one: *if* the future will be sufficiently good. It might not be. Let's look at this in the next chapter.

CHAPTER 9

Will the Future Be Good or Bad?

Sentience as a Single Life

In the opening of this book I asked you to consider humanity as a single life, where you live every human life that has ever been lived, reincarnated into one after the other. Let's return to this thought experiment and ask some further questions. First, has it all been worth it? If you lived through every life up until today, would you think that your life has been good, on balance? Are you glad that you lived those hundred billion lives? Second, when you look to the future, is it with a sense of optimism or dread? If you found out that the human race was certain to peter out within the next few centuries, would you greet that knowledge with sadness because of all the joys you would lose or with a sense of relief because of all the horrors you would avoid?

And let's reflect on how our answers to these questions might change if we altered the thought experiment. Rather than living through just the hundred billion human lives that have existed to date, imagine that instead you live through the lives of all sentient creatures.[1] The first invertebrate brains evolved over five hundred million years ago;[2] we don't know when the first flame of consciousness was kindled—that is, when the first *experience* occurred—but it might have been not so long after. For this thought experiment, however, let us make the conservative assumption that only vertebrates are sentient. If you lived through the lives of all conscious beings, you would then experience a hundred billion trillion years of sentience. You would spend nearly 80 percent of your time as a fish. You would spend 20 percent of your time—thirty billion trillion years—as an amphibian or reptile. You would spend one quadrillion years living as various kinds of dinosaurs before dying because of an asteroid impact in the last mass extinction. Your time as a mammal would make up only one-thousandth of your existence.[3]

Your life as a human being would amount to only one–hundred billionth of your time on Earth. If this were your life, the evolution of *Homo sapiens* would be a jarring event: for the first time you would no longer merely be experiencing; you would also be able to understand and conceptualize your experiences. During this time, the natural environments you had been living in would be progressively destroyed, and you would find yourself experiencing, for the first time, the many lives of animals bred and slaughtered for human consumption. If you were living through the lives of all sentient beings, would you regard the evolution of *Homo sapiens* as a good thing? And, looking ahead, if you knew you were going to experience all future sentient lives, including those of any sentient artificial beings that might one day be created, would you feel optimistic?

This thought experiment sets the stage for the question that this chapter addresses: Should we expect the continuation of civilisation into the distant future to be a good thing, morally speaking? Or should we think that if civilisation were to end in the next few centuries, the world would be better off for it? This is a crucial question for longtermists because it affects how we should prioritise among our efforts. Let's call those who think that the prospective future is good *optimists* and those who think that the prospective future is bad *pessimists*. The more optimistic we are, the more important it is to avoid permanent collapse or extinction; the less optimistic we are, the stronger the case for focusing instead on improving values or other trajectory changes.

Philosophers have been divided on this question of how optimistic or pessimistic we should be about the future. The notoriously dour Schopenhauer, for example, suggested that "it would have been much better if the sun had been able to call up the phenomenon of life as little on the earth as on the moon; and if, here as there, the surface were still in crystalline condition."[4] More prosaically, David Benatar recently claimed that "although the prospect of human extinction may, in some ways, be bad for us, it would be better, all things considered, if there were no more people (and indeed no more conscious life)."[5]

In contrast, in his last work, *On What Matters*, Parfit took an optimistic stance, commenting,

> Just as we had ancestors who were not human, we may have descendants who will not be human. We can call such people *supra-human*. Our descendants might, I believe, make the further future very good. . . . Life can be wonderful as well as terrible, and we shall increasingly have the power to make life good. Since human history may be only just beginning, we can expect that future humans, or supra humans, may achieve some great goods that we can not now even imagine. In Nietzsche's words, there has never been such a new dawn and clear horizon, and such an open sea.[6]

The question of the value of the future is tricky, but I'll suggest that, all things considered, we should expect the future to be positive on balance. I'll first discuss whether the world is good on balance for people alive today and whether it's getting better or worse; I'll then do the same for nonhuman animals and for what philosophers call "non-welfarist goods." Finally, I'll discuss how we should weigh up goods against bads and give an argument for optimism about the longterm future.

How Many People Have Positive Wellbeing?

Let's start our investigation into the value of the future by asking whether right now, the world is better than nothing for the human beings alive today. Do most people have lives that are positive, on balance? This topic is a sensitive and difficult one, but it seems to be possible for people to have lives of negative-wellbeing. If someone's life consists only of intense suffering and torture, it clearly makes sense to say that their life is bad for them. As I emphasised last chapter, this is not to say that their lives are "not worth living"—someone could have a life such that they would prefer to have never been born and yet contribute enormously to society through their work and their relationships. Rather, it is to say that, from that person's perspective, putting to the side any effects on others, their life involves so much suffering that it is worse than nonexistence.

You might ask, Who am I to judge what lives are above or below neutral? The sentiment here is a good one. We should be extremely cautious in trying to figure out how good or bad others' lives are, as it's so hard to understand the experiences of people with lives very different to one's own. The answer is to rely primarily on people's self-reports. As we'll see, the best evidence regarding how many people in the world today have lives that are below neutral comes from simply asking people to say, in their own view, whether their lives contain more suffering than happiness, or whether they would prefer to have never been born.

The question of how many people have lives of positive wellbeing—and what it is that makes their lives good—is not just important for longtermists. It's also relevant, for example, for governments deciding how to prioritise health-care resources. If you think that most people have only slightly positive wellbeing, then you will be more inclined to favour funding interventions that improve lives, such as treating chronic pain, over policies that save lives, such as preventing malaria; if you think that most people have great lives, then saving lives becomes comparatively more important. Remarkably, the leading approach to measuring the burden of disease, which is widely used by governments and philanthropists when setting health-care policy, assumes that death is the worst possible state one can be in, even though this is clearly false.[7] It thereby systematically biases policies towards saving life over improving quality of life.

You might think it's obvious that the vast majority of people have lives with net positive wellbeing. I certainly think I have such a life, and you might feel the same. But I am extremely unrepresentative of the world as a whole, and if you're reading this book, you probably are, too. More than half the people in the world live on less than seven dollars per day, and that figure already accounts for the fact that money goes so much further in poor countries: it represents the equivalent of what seven dollars would buy in the United States.[8] I would not, intuitively, regard myself as exceptionally wealthy; I live on an income that's only a little higher than the median income in the UK. But even given this, I'm a full fifteen times richer than the majority of people in the world.[9] I therefore shouldn't expect to be able to imagine the life and wellbeing of the average person alive today, let alone the poorest billion people alive.

In order to assess whether most people have net positive wellbeing, the first thing we need to be clear about is what wellbeing *is*. In moral philosophy, there are three main theories of wellbeing.[10] The first is the preference satisfaction view, according to which your life goes well to the degree to which your carefully considered preferences about your life are fulfilled. In this view, your life going well is about getting what you want, even if that does not impact your conscious experiences in any way. For example, you might have a preference for your partner to be faithful to you, even in situations where you would never know either way.

The second view is hedonism, according to which your wellbeing is entirely determined by your conscious experiences: positive experiences, like pleasure or tranquillity, make your life better, while negative experiences, like pain or sadness, make your life worse. In this view, getting what you want does not make your life better unless it improves the balance of positive and negative conscious experiences. If someone wants to become rich and succeeds, but they have just the same balance of negative and positive experiences as before, the hedonist would say that this person's life has not improved merely by virtue of getting what they wanted.

The third view is what's called the "objective list" view. In this view, there are many things that can improve your wellbeing even if they do not improve your conscious experiences and even if you don't desire them. This is why they are called "objective" goods. These could include things like friendship, the appreciation of beauty, or knowledge. The questions I address in this chapter are particularly hard to assess in the objective list view—not least because there is such a diversity of objective goods—so I have to put them to the side, although I have a section on non-wellbeing goods which will help shed some light on the issue.

Unfortunately, despite the importance of the issue of how many people have net positive lives, the psychological data we have on it is extremely limited. Out of 170,000 books and papers published on subjective wellbeing,[11] only a handful have directly addressed the question of for whom life is positive on balance. There are three main psychological approaches that bear on this issue.[12]

First are surveys that try to measure people's life satisfaction. Life satisfaction surveys ask respondents to rate their lives, as a whole, on a scale from 0 to 10, where 10 represents the best possible life for them and 0 represents the worst possible life for them.[13] Survey data of more than 1.5 million people from 166 countries found that, from 2005 to 2015, only 47 percent of respondents had mean scores above 5.[14]

For our purposes, though, what we need to know is how survey respondents interpret the scale and, in particular, where the neutral point is—the point on the scale below which they think life is so bad that it's worse, for them personally, than being dead. We can't assume that this is the midpoint of the scale. Indeed, it's clear that respondents aren't interpreting the question literally. The best possible life (a 10) for me would be one of constant perfect bliss; the worst possible life (a 0) for me would be one of the most excruciating torture. Compared to these two extremes, perhaps my life, and the lives of everyone today, might vary between 4.9 and 5.1.[15] But, when asked, people tend to spread their scores across the whole range, often giving 10s or 0s. This suggests that people are relativising their answers to what is realistically attainable in their country or the world at present.[16] A study from 2016 found that respondents who gave themselves a 10 out of 10 would often report significant life issues. One 10-out-of-10 respondent mentioned that they had an aortic aneurysm, had had no relationship with their father since his return from prison, had had to take care of their mother until her death, and had been in a horrible marriage for seventeen years.[17]

The relative nature of the scale means that it is difficult to interpret where the neutral point should be, and unfortunately, there have been only two small studies directly addressing this question. Respondents from Ghana and Kenya put the neutral point at 0.6, while one British study places it between 1 and 2.[18] It is difficult to know how other respondents might interpret the neutral point. If we take the UK survey on the neutral point at face value, then between 5 and 10 percent of people in the world have lives that are below neutral.[19] All in all, although they provide by far the most comprehensive data on life satisfaction, life satisfaction surveys mainly provide insights into *relative* levels of wellbeing across different people, countries, and demographics. They do not provide much guidance on people's *absolute* level of wellbeing.

A second line of evidence is from surveys that simply ask people if they are happy. The World Values Survey asks respondents whether they are "very happy," "rather happy," "not very happy," or "not at all happy." The last survey was in 2014 and included respondents from sixty countries, comprising

67 percent of the world population. It found that in all countries except Egypt (which was undergoing a protracted political crisis at the time), more than half of people rate themselves as very happy or rather happy, and in almost all countries, more than 70 percent of people say they are happy.[20] In several countries, reported rates of happiness are extremely high. In Qatar, 98 percent of people say they are happy, as do 95 percent of Swedes, and 91 percent of Americans. Even in a poor country like Rwanda, 90 percent of people say they are happy.

These ratings are probably overly optimistic.[21] For example, in 2013, one survey found that 11 percent of Swedish adults were experiencing clinical depression at a particular point in time, but in the World Values Survey, only 5 percent of Swedes rated themselves as unhappy.[22]

The third line of evidence on whether people have lives with positive wellbeing comes from early and intriguing work using an experiential approach to measuring wellbeing: asking people at random times how they feel in that moment. This is known as "experience sampling." Those who favour this method of measuring happiness argue that it avoids some of the biases inherent in the life satisfaction approach, such as that people might have a selective memory, or that questions about life satisfaction measure people's perceptions of their own social status rather than their happiness.

In a currently unpublished large survey of over 8,500 people, psychologists Matt Killingsworth, Lisa Stewart, and Joshua Greene added a twist to the experience sampling approach.[23] At random times, they asked participants to write down what activity they were doing and how long it would last, and then respond to the question, "If you could, and it had no negative consequences, would you jump forward in time to the end of what you're currently doing?" That is, they asked participants to imagine having the option of simply not *experiencing*—though still doing—whatever activity they were engaged in at that moment. If they were making a cup of tea, they would imagine that they could blink and their next experience would be drinking the cup of tea that they had just made. The researchers called this "skipping" an experience. The idea underlying the question was that, if someone would choose to skip an experience, they were judging that experience to be worse than nothing; if someone chose to keep an experience, they were judging that experience to be better than nothing.

It turns out that people in the survey, on average, would skip around 40 percent of their day if they could. In a second, smaller study, the same experimenters asked people to look back at the previous day and indicate which experiences they would have skipped if they could, and then asked them to compare pairs of experiences with each other to work out how good the experiences they'd have kept were and how bad the experiences they would've skipped were. For instance, a study subject might say that thirty minutes of an activity they'd rather skip—say, housework—was worth fifteen minutes of an enjoyable activity—say, dinner with friends. This would indicate that, for this study subject, having dinner with friends is twice as good per minute as doing housework is bad. Again, people skipped around 40 percent of their day, and on average, people were happier during the times they kept than they were unhappy during the times they skipped. Taking both duration and intensity into account, the negative experiences were only bad enough to cancel out 58 percent of people's positive experiences.

The sorts of experiences people kept and skipped were what you might expect: people skipped 69 percent of the time they were working and only 2 percent of the time they were engaged in what the experimenters euphemistically called "intimate activities." In the smaller of the two studies, in which intensity of experience was measured, 12 percent of people had lives where, on the day in question, negative experiences outweighed the positive. This does not necessarily mean that 12 percent of people have lives of negative wellbeing—these respondents might just have had a bad day.[24]

The results of these studies might seem like positive news, seeing as the participants in the study had good lives on average. But I think the right conclusion is actually more pessimistic.[25] The participants in these studies mainly lived in the United States or in other countries with comparatively high income levels and levels of happiness, and the ones in the larger study all owned an iPhone. In 2016, Apple was the consumer brand that best predicted whether a purchaser was rich and well educated (in 1992, the brand that best predicted income was Grey Poupon mustard).[26] The skipping studies were therefore somewhat skewed towards wealthier and better-educated people, and the results were not representative of the lives of prisoners, who in the United States constitute 0.7 percent of the population, or the homeless (0.17 percent of the US population). Yet, even within such a selected sample, participants said they would choose to skip 40 percent of their life, their bad experiences cancelled out nearly 60 percent of their good experiences, and, for more than a tenth of people, negative experiences outweighed the positive. Overall, while this study is highly intriguing and well done, it's limited in what it tells us about global happiness.

Because the published evidence we have so far is so limited, I commissioned three psychologists—Lucius Caviola, Abigail Novick Hoskin, and Joshua Lewis—to run a survey on the topic.[27] They asked 240 people in the United States and 240 people in India a range of questions on the quality of their life so far, including these:

> Do you think that your life to date has involved more happiness than suffering?
>
> Ignoring any effects of your life on other people, would you prefer to be alive or would you prefer to have never been born?
>
> If you could live the exact same life again from the beginning (without remembering anything from before, so you would experience everything as if for the first time), would you do it? Assume this decision affects no one else and you are just deciding for your own benefit.

They also asked for qualitative comments. One respondent simply said, "Those are some deep questions, man." Those who gave positive answers often wrote quite beautiful responses, such as, "I'm happy I was born to experience so many things such as the births of my nieces and nephews and many children I have watched grow. . . . I also love the wonder of it all, the birds, butterflies, trees, rivers are all so beautiful."

The comments from those who gave negative answers were as dark as one might expect, such as, "My life was and is a horrible thing. I would not want to relive it again," and "I have lived through pure hell the last 20 years of my life and I would not wish it on anyone."

Positive answers were much more common than negative answers. In the United States, 16 percent said that their life contained more suffering than happiness, and 40 percent said it contained more happiness than suffering.

Nine percent preferred never to have been born, and 79 percent preferred to be alive. Thirty percent would not live the exact same life again, and 44 percent would.

The results were similar in India, although, strikingly, Indian respondents were more positive than those from the United States. Only 11 percent thought their lives contained more suffering than happiness, only 6 percent preferred never to have been born, and only 19 percent would choose not to live their life again. This might be simply because the samples were not representative of the population as a whole: respondents tended to be comparatively well-off Indians and comparatively less well-off Americans.

Table 9.1. How Many People Live Lives of Positive Wellbeing? Evidence from a Survey in India and the United States (in percent)

	India			United States		
Question	Negative	Neutral	Positive	Negative	Neutral	Positive
Do you think that your life to date has involved more happiness than suffering?	11	52	37	16	44	40
Ignoring any effects of your life on other people, would you prefer to be alive (instead of to have never been born)?	6.3	8.4	85	9.1	13	79
If you could live the exact same life again from the beginning (without remembering anything from before, so experiencing everything as if for the first time), would you do it? Assume this decision affects no one else, and you are just deciding for your own benefit.	19	12	69	31	25	44

Notes: Data from Caviola et al. 2021. Percentages might not sum to 100 because of rounding.

How should we put this all together? The conclusions we come to will vary depending on the theory of wellbeing we invoke. Life satisfaction scores, in which people rate their own happiness, seem to more closely track a preference satisfaction view, since people saying they are satisfied with their life is evidence that their preferences are being satisfied. The skipping studies more closely track a hedonistic view of welfare: even if people desire to be in their job, for instance, the evidence suggests that many of them do not enjoy it very much, and the skipping study captures that fact. The World Values Survey, which directly asks people whether they are happy, is perhaps most naturally interpreted in a preference-satisfactionist way, but one could imagine that some respondents also interpreted it in a hedonist way.

I would tentatively suggest that something like 10 percent of the global population have lives with below-neutral wellbeing. If we assume, following the small UK survey, that the neutral point on a life satisfaction scale is between 1 and 2, then 5 to 10 percent of the global population have lives of negative wellbeing. In the World Values Survey, 17 percent of respondents classed themselves as unhappy. In the smaller skipping study of people in rich countries, 12 percent of people had days where their bad experiences outweighed the good. And in the study that I commissioned, fewer than 10 percent of people in both the United States and India said they wished they had never been born, and a little over 10 percent said that their lives contained more suffering than happiness.

So, I would guess that on either preference-satisfactionism or hedonism, most people have lives with positive wellbeing. If I were given the option, on my deathbed, to be reincarnated as a randomly selected person alive today, I would choose to do so. If I were to live through the lives of everyone alive today, I would be glad to have lived.

Next, let's ask how human wellbeing is changing over time. Are people getting happier or staying much the same?

Are People Getting Happier?

A common view is that, even though the world is getting richer, people are no happier or are even becoming less happy. In support of this view, one could point to the Easterlin paradox: although higher income is correlated with greater happiness both within and across countries at a specific point in time, *over time*, people and countries do not get happier as they get richer.[28] In this view, it's relative income within a country that determines a person's happiness; our absolute level of income is irrelevant because we get accustomed to whatever level of income we have. In this view, then, insofar as income inequality within countries is generally increasing over time, we might expect people to get less happy over time.

However, though Easterlin's paradox continues to be influential, it doesn't actually exist. Easterlin first published his findings back in 1974, when the data we had about levels of happiness around the world was much more sparse than it is today.[29] From the fact that we could not, at the time, show that countries get happier as they get richer, he concluded that there was no relationship between absolute level of income and happiness and that happiness was instead determined by one's income relative to one's peers.[30] But more recent work with better data strongly supports the view that countries get happier as they get richer.[31] It may well be that your relative level of income within your country influences how happy you are, but it's also true that your happiness increases with your absolute level of income.

Figure 9.1 shows the average happiness of a country compared to its GDP per person.[32]

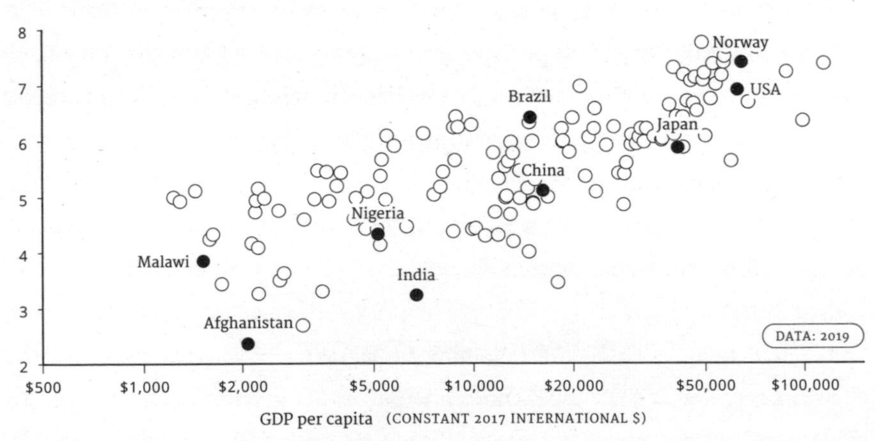

Figure 9.1. Self-reported life satisfaction (on a scale from 0 to 10) vs. per-capita income (adjusted for price differences between countries). Each circle represents one country.

And Figure 9.2 is the graph of the happiness of countries over time, as they've gotten richer.

Figure 9.2. As countries become richer, their population tends to become happier, as evident from the fact that most arrows point towards the top right. This is true worldwide—for the full data, see Figure Credits and Data Sources, page 263.

Even though richer people tend to be happier, it is not clear whether this effect is causal. Maybe happier people are easier to work with and so tend to earn more money. One way to explore the causal effect of money on happiness is by looking at lottery winners. Newspapers and magazines often report about the so-called curse of the lottery, of newly minted but miserable millionaires. In 2016, *Time* magazine published a piece called "Here's How Winning the Lottery Makes You Miserable," with several anecdotes of people whose lives had been ruined by fabulous wealth.[33] The only exception mentioned in the article was Richard Lustig, who won substantial lottery prizes no fewer than seven times and wrote the book *Learn How to Increase Your Chances of Winning the Lottery*.[34] Lustig said, "I've been rich and I've been poor, and I like rich a whole lot better." It turns out that Lustig's experience is actually more representative of lottery winners as a whole. Recent research has found that lottery winners are happier.[35] This is further evidence for the view that money does improve people's wellbeing.

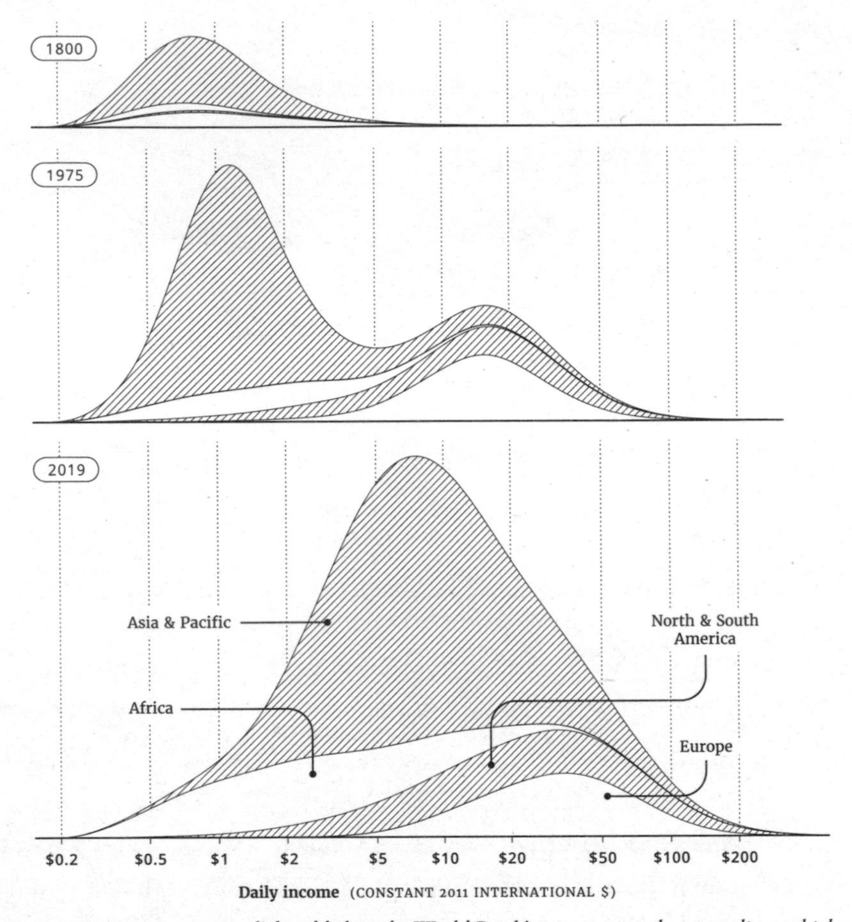

Figure 9.3. In 1800, most people lived below the World Bank's international poverty line, which indicates extreme poverty ($1.90 per day). By 1975, a group of rich countries—mostly in Europe and the Americas—had pulled away and enjoyed historically unprecedented per-capita incomes. Forty years later, while stark global inequalities remain, the overall income distribution shows less polarisation between rich and poor people, and an increasing share of the population—particularly in Asia—has escaped extreme poverty. All income figures are adjusted for inflation and price differences between countries.

The literature on subjective wellbeing is generally supported by other measures of how well-off people are, on average. For example, Figure 9.3 shows how the global distribution of income has changed over time.[36]

And Figure 9.4 shows life expectancy at birth for the world as a whole and for the six most populous low- and middle-income countries.

One study found that in countries experiencing sustained economic growth, happiness inequality has been decreasing over time, even in

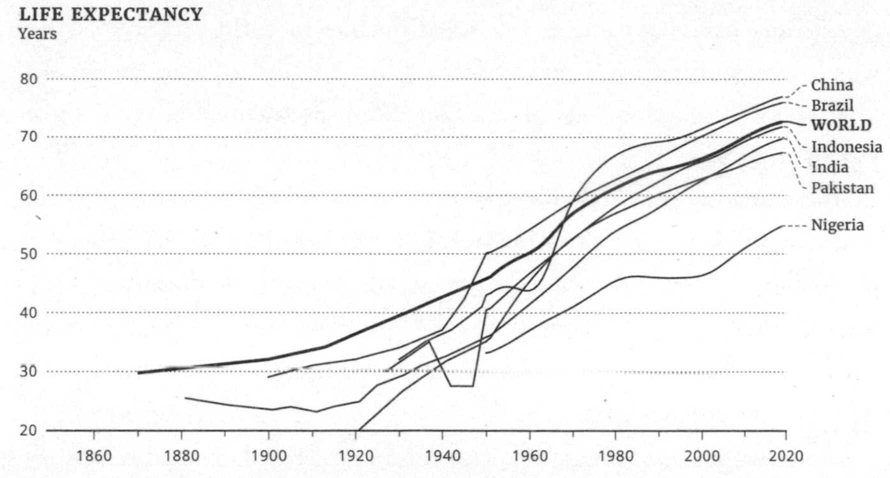

Figure 9.4. Life expectancy has more than doubled in many countries since the nineteenth century. Both for the world as a whole and for the six most populous low- and middle-income countries, it has increased almost every year for decades.

countries which have also experienced rising income inequality.[37] This is true across socioeconomic classes and across different races. The authors of the study suggest that as countries get richer, their governments spend more on things like health, infrastructure, and social protection, which affect incomes and happiness differently.

Similarly, in the United States the Black-White happiness gap has closed by two-thirds since the 1970s, although today White Americans remain happier on average, even after controlling for differences in education and income. Inequality between self-reported happiness scores has also decreased between genders. But this might not be for the reason you think: surprisingly, it's because women have gotten less happy over time. They used to report being happier than men, but now they are similar in happiness to men. It's not currently known why this trend has occurred.[38]

These broad improvements in human wellbeing are an important corrective to the widespread belief that the world is getting worse and will continue to do so. While some people may be steadfast optimists, there is a lot of evidence that many of us are pessimistic about how the rest of the world is doing—arguably too pessimistic.[39] A 2015 survey of eighteen thousand adults found that in many rich countries, less than 10 percent of respondents think the world is getting better.[40] This pessimism is driven in part by the negative skew of news. A huge plane crash makes for compelling news, but a long sustained decline in child mortality is not worth mentioning: if it bleeds, it leads. This leads us to focus on the bad and ignore the good, so we miss the huge improvements that are happening all around us.

These trends, though, do not give reasons for thinking that the problems in the world today are not so bad after all. I mentioned earlier that most people still live on less than seven dollars per day; in addition, every year millions die from easily preventable diseases, millions more are oppressed and abused, and hundreds of millions of people go hungry. This is not a world we should be content with.

Moreover, average human wellbeing has not increased inexorably upwards throughout all of human history. While living standards today are undoubtedly much higher than they were in preindustrial agricultural societies, our nomadic hunter-gatherer ancestors, from the dawn of *Homo sapiens* up until the agricultural revolution around ten thousand years ago, probably had higher average wellbeing than early agriculturalists. As people relied more on agriculture, their height—a good indicator of nutrition and health—usually declined compared to that of their hunter-gatherer ancestors.[41]

There is even some evidence that, in some ways, the lives of preagricultural hunter-gatherers were quite attractive compared to the life of the average person alive today.[42] Although measurement is difficult, on average, the working hours of modern hunter-gatherers are not dramatically different from those in modern industrial societies, and for some hunter-gatherers they are much lower.[43] Moreover, many hunter-gatherers enjoy their work—after all, hunting is a popular recreational activity for many people today. Hunter-gatherers usually have a strong egalitarian ethos and high levels of community,[44] and they dance and sing regularly. In his study of the Hadza from Tanzania, one of the few remaining hunter-gatherer societies in the world, the anthropologist Frank Marlowe noted,

> The Hadza sing often, and everyone can sing very well. When several Hadza get in my Land Rover to go somewhere, they almost invariably begin singing. They use a melody they all know but make up lyrics on the spot. These lyrics may go something like "*Here we go riding in Frankie's car, riding here and there in the car. When Frankie comes, we go riding in the car.*" They take

> different parts in a three-part harmony, never missing a beat, all seemingly receiving the improvised lyrics telepathically.
>
> They also love to dance and do so in various distinct styles. . . . This dancing is unique and full of soul—the most sensual dancing I've ever seen.[45]

The Hadza were involved in the only study that has compared wellbeing in a hunter-gatherer group whose life might be comparable to our distant ancestors with wellbeing in industrialised nations. Although their diet is generally pretty good, the Hadza otherwise subsist in material poverty: they own few possessions and live in temporary shelters made of dried grass and branches.[46] Despite that, the study found that the Hadza people were happier than all twelve industrialised populations for which comparable scores are available.[47] We should be careful when drawing conclusions about preagricultural hunter-gatherers from modern hunter-gatherers because modern hunter-gatherers are different in several important respects: they live at environmental extremes, and they have conflicts with and trade with modern societies.[48] Moreover, hunter-gatherer lifestyles vary widely, and the Hadza are especially harmonious,[49] so they may not be representative, and this is only one study. The evidence is intriguing nonetheless. Perhaps the strongest evidence on hunter-gatherer happiness is qualitative. Ethnographers regularly comment on the apparent harmony and desirability of the hunter-gatherer lifestyle.[50]

A great drawback of being a preindustrial hunter-gatherer was that, because of disease, occasional hunger, and the lack of modern medicine, life expectancy was much lower than it is today (though higher than in early agricultural societies). Around half of children born in preagricultural hunter-gatherer societies died before the age of fifteen, compared to one in two hundred in Europe today.[51] If a hunter-gatherer made it to age fifteen, they could expect to live until fifty-three, whereas the average Brit who makes it to fifteen today can expect to live until eighty-nine.[52] Some scholars also argue that rates of violence were much higher among preagricultural hunter-gatherers, though this is fiercely disputed.[53]

Since the Industrial Revolution, there has been a clear upward trend in wellbeing, and this gives us good reason to believe that the world will continue to get better for people over at least the next century. On most economic forecasts, the world will continue to grow richer over the coming decades. Over the last fifty years, global GDP per person grew by 2 percent per year, and all major geographic areas are experiencing significant economic growth.[54] In one recent survey of growth economists, the respondents thought that this trend would stay broadly the same, at 2.1 percent per year;[55] given this, by 2100 the average person will be five times richer than they are today and so probably will also be happier. Over the course of the next century, at least, we have grounds for optimism.

Nonhuman Animals

So far, we've just looked at whether the average human life is better than nothing. But in order to assess whether the world as a whole is good on balance and whether it's getting better, we need to look more widely than this. In particular, we've not yet discussed the vast majority of sentient beings on this planet: nonhuman animals. We'll start with farmed animals.[56]

As of 2018, there were more than 79 billion vertebrate land animals killed for food every year; of these, there were 69 billion adult chickens, 3 billion baby male chicks, 3 billion ducks, 1.5 billion quail, 1.5 billion pigs, 922 million rabbits and hares, 656 million turkeys, 574 million sheep, 479 million goats, and 302 million cattle. In addition, around 100 billion fish are slaughtered in fish farms every year.[57]

The suffering we inflict on these animals is difficult to overstate.[58] Chickens, who make up the vast majority of land animals killed for food, probably suffer most. Chickens raised for meat, called broiler chickens, are bred to grow so quickly that by the end of their life, 30 percent have moderate to severe walking problems. When they're big enough to be slaughtered, most broiler chickens are hung upside down by their legs, their heads are passed through electrified water, and then, finally, their throats are cut. Millions of chickens survive this only to finally die when they are submerged in scalding water in a step of the process meant to loosen their feathers.[59]

Egg-laying chickens likely suffer even more, starting the moment they hatch. Male chicks are useless to the egg industry and are therefore "culled" as soon as they're born. They're either gassed, ground up, or thrown into the garbage, where they either die of thirst or suffocate to death. But compared to the suffering that awaits female chicks, the culled male chicks may be the fortunate ones. Once grown, many hens are confined to battery cages smaller than a letter-size piece of paper. Egg-laying hens are prone to peck other hens, which in some cases ends up in cannibalism. To prevent this, a hot blade or infrared light is used to slice off the tips of female chicks' extremely sensitive beaks. After enduring mutilation as chicks and intense confinement as adults, many egg-laying hens nearing the end of their productive lives are subjected to forced molting: they are starved for two weeks, until they lose a quarter of their body weight, at which point their bodies start another egg-laying cycle. Once they become so unproductive as to be unprofitable, they are gassed or sent to a slaughterhouse.

Farmed cattle and pigs have better lives than this, but they still suffer much unnecessary pain. Pigs are castrated and have their tails amputated, and farmed cattle are castrated, dehorned, and branded with a hot iron—all without anaesthetic. Female pigs and dairy cows endure artificial insemination, which is painful and invasive, at least once a year. After that, things only get worse for them. During pregnancy, the overwhelming majority of female pigs are confined to gestation crates so small they can't turn around. Female cows in industrial farms are subjected to mechanized milking for ten out of twelve months of the year, before they're "spent" and slaughtered at around five years old. Their male calves, of no use to the dairy industry, are sold to veal factories, where they're kept in tiny stalls and, in many countries, tethered to the wall for the entirety of their short lives.[60]

Farmed fish also suffer terribly. Fish farms are very overcrowded: salmon, which are around seventy-five centimetres long, can be given the space equivalent of just a bathtub of water each.[61] This overcrowding precludes natural behaviour and leads to injury and premature death. Mortality in fish farms ranges from 15 percent to 80 percent.[62] Atlantic salmon and rainbow trout are starved for several days, sometimes for two weeks or more, to empty the gut before slaughter.[63] Most farmed fish are killed by being left to asphyxiate slowly to death, which can take more than an hour.[64] Others are gassed with carbon dioxide or have their gills cut while still conscious.[65]

Putting this all together, it seems hard to resist the conclusion that, when a factory-farmed chicken, pig, or fish dies, that's the best thing that's happened to them. I know of few people who've studied the issue intensively and disagree.[66] In totality, industrial farming consists in the efficient, society-wide production of a monstrous volume of suffering.

The question of what weight to give to human interests and to nonhuman animal interests is difficult.[67] Humans are literally outweighed by farmed animals: land-based farmed animals have 70 percent more biomass than all humans.[68] Land-based farmed animals also outnumber humans greatly, by a factor of three to one, with 25 billion chickens, 1.5 billion cattle, 1 billion sheep, and 1 billion pigs alive at any one time; farmed fish outnumber us, at a very rough estimate, ten to one, with around 100 billion farmed fish alive at any one time. However, these species do not all have equal capacity for wellbeing, and it's hard to believe that capacity for wellbeing does not matter at all when comparing the interests of different species. Accounting for differences in capacity for wellbeing does not entail that other species have lower moral status than humans. Rather, it gives their wellbeing equal weight but recognises that some species simply have less of it than others.

To capture the importance of differences in capacity for wellbeing, we could, as a very rough heuristic, weight animals' interests by the number of neurons they have. The motivating thought behind weighting by neurons is that, since we know that conscious experience of pain is the result of activity in certain neurons in the brain, then it should not matter more that the neurons are divided up among four hundred chickens rather than present in one human. If we do this, then a beetle with 50,000 neurons would have very little capacity for wellbeing; honeybees, with 960,000 neurons, would count a little more; chickens, with 200 million neurons, count a lot more; and humans, with over 80 billion neurons, count the most.[69] This gives a very different picture than looking solely at numbers of animals: by neuron count, humans outweigh all farmed animals (including farmed fish) by a factor of thirty to one. This was very surprising to me; before looking into this, I hadn't appreciated just how great the difference in brain size is between human beings and nonhuman animals.

If, however, we allow neuron count as a rough proxy, we get the conclusion that the total weighted interests of farm land animals are fairly small compared to that of humans, though their wellbeing is decisively negative.

This does not yet resolve whether the welfare of humans and farmed animals combined is negative. Even though, in totality, farmed animals may have fewer neurons, the vast majority of farmed animals (chicken and fish) live lives full of intense suffering, which could well outweigh total human wellbeing. If the intensity of the suffering of chickens and fish is at least forty times the intensity of average human happiness, then the combined wellbeing of humans and farmed animals is negative.

Next, we can turn to assessing the lives of animals in the wild. When we try to weigh the wellbeing of wild animals by their number or neuron count, we get the conclusion that our overall views should be almost entirely driven by our views on fish.[70] The biomass of human beings is five times larger than the biomass of all wild birds, reptiles, and mammals combined,[71] and humans have three times as many neurons. But the biomass of fish is ten times larger than that of humans,[72] and there are at least ten thousand times as many fish as human beings. Most of these are tiny fish weighing a few grams that live two hundred to a thousand metres below the ocean surface.[73] Although these fish each only have around twenty million neurons,[74] conservative calculations suggest that, by neuron count, fish outweigh humans by at least a factor of seventeen.

Table 9.2. Counting Individual Animals vs. Counting Neurons

Species	Total population	Total neurons
Humans	8 billion	700 million trillion
Farmed animals	135 billion	20 million trillion
Wild fish	600 trillion	12 billion trillion

Notes: Population data from FAOSTAT and Carlier and Treich (2020). Neuron counts based on Olkowicz et al. (2016, Table S1); Herculano-Houzel et al. (2015); and Herculano-Houzel (2016, 75). Details and bibliographic information available at whatweowethefuture.com/notes.

How good is the life of a wild fish? It's easy to have a rosy picture of life in the wild as being in pleasant harmony with the earth, but Tennyson's line about "Nature, red in tooth and claw" is more accurate. While some adult fish species can live for decades, more than 90 percent of fish larvae die mere days after hatching—eaten, starved, or suffocated.[75] Those that make it to adulthood may suffer from diseases—fungal, bacterial, and viral infections—just as humans do. And the vast majority of adult fish will die not of old age but will instead suffocate as a result of an algal bloom, or be killed by parasites, or die of exhaustion after building their nest or releasing their eggs, or be torn apart or swallowed whole then crushed in a predator's oesophagus.[76]

It's common to think of the experiences of wild animals as part of a "circle of life" that is at best a miracle of nature and at worst just part of the natural order. But while many people ignore the pain of animals in the wild, we feel sympathy and even outrage when animals are injured or killed as a result of human intervention. In my view, there's no good reason for this: a turtle ripped apart by a killer whale experiences no less pain than one strangled by the plastic loops that held together a six-pack.[77]

Overall, do the lives of wild animals involve more suffering than joy? Put another way: If you were given the option, on your deathbed, to be reincarnated as a randomly selected animal in the wild, would you do so? I really don't know if I would. It's very hard to make confident inferences about the wellbeing of wild animals, given that their physiologies and lives are so

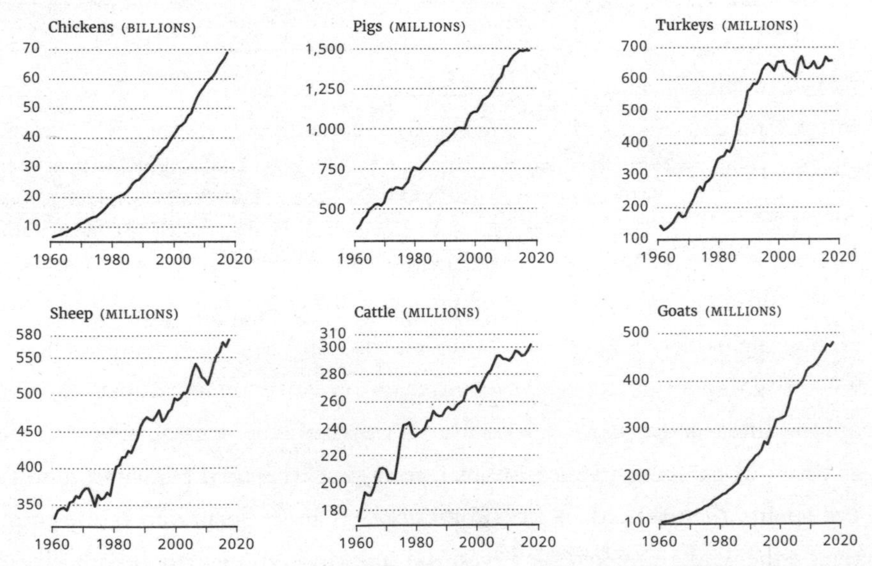

Figure 9.5. Humans have been slaughtering ever-larger numbers of farmed land animals over the last six decades. Data exclude egg and dairy production.

different from our own. Overall, it's at best highly unclear, given what we currently know, whether wild animals have positive wellbeing or not.[78]

Our overall assessment of the lives of animals is therefore fairly pessimistic. Farmed animals probably have lives of negative wellbeing, on average. For wild animals it's unclear, but their lives may well be negative on average, too. Next, we should ask, Are the lives of nonhuman animals improving over time or getting worse?

The trend is clearly negative for farmed animals. The number of animals raised for food is growing very rapidly, with consumption increasing fastest among chickens and pigs, which, as we've seen, have among the worst lives.

The wellbeing of animals raised for consumption is getting worse over time, too, as we develop ever more "efficient" methods for turning feed into meat. In particular, selective breeding means that modern chickens now grow unnaturally quickly and to unnaturally large sizes; this means that they suffer from a range of skeletal disorders and deformities, are often crippled later in life, and may be chronically hungry because of food restrictions.[79] Some countries have improved their animal welfare laws, but that is a small effect compared to these other factors. In spite of all of this, our *attitudes* towards animals have clearly improved over the last few hundred years, which could provide some hope for the future.

The trend for wild animals is less clear. Human expansion means that the biomass of wild land mammals has decreased by a factor of seven compared to prehuman times, mostly due to the megafaunal extinctions I discussed in Chapter 2.[80] The biomass of commercially caught predatory fish has declined dramatically, but this has been to some extent offset by an increase in the biomass of smaller prey fish.[81] On balance, various studies suggest that human activity over the last forty years has probably decreased vertebrate and invertebrate populations, though the evidence is limited and somewhat conflicting.[82] How you evaluate this depends on your view on wild animal wellbeing. It's very natural and intuitive to think of humans' impact on wild animal life as a great moral loss. But if we assess the lives of wild animals as being worse than nothing on average, which I think is plausible (though uncertain), then we arrive at the dizzying conclusion that from the perspective of the wild animals themselves, the enormous growth and expansion of *Homo sapiens* has been a good thing.

Non-wellbeing Goods

So far we've looked at trends in the wellbeing of both human beings and nonhuman animals.

You might think that wellbeing is all that matters, morally. This is the view that, after philosophical reflection, I find most plausible: other things can be valuable or disvaluable *instrumentally*, but only insofar as they ultimately impact the wellbeing of sentient creatures. But philosophers are split on this issue: many would reject the idea that only wellbeing is of moral value and claim that there are things that can make the world better or worse even if they are not good or bad *for* any sentient creature. For example, philosopher G. E. Moore claimed that natural or artistic beauty is good regardless of whether people appreciate it; many environmentalists think that ecosystems being allowed to run their natural course is a good thing in and of itself, irrespective of the wellbeing of the individual animals that live and die in those ecosystems.[83]

Given the difficulty of ethics and, as I argued in the last chapter, the need for us to acknowledge moral uncertainty, we should consider the trend in non-wellbeing goods. Unfortunately, it is hard to make robust arguments that establish what things are valuable over and above their effect on wellbeing; this is an area of ethics where we may be able to do no better than have our intuitions about fundamental values butt against each other. Some possibilities that many people find compelling, in addition to great art and the natural environment, are democracy, equality, the spread of knowledge, and great human accomplishments.

It's not clear whether the trend of non-wellbeing goods has increased or decreased over time. In terms of the natural environment, the trend looks negative. We have cut down one-third of the world's forests. Global forest area continues to decline, but there is some cause for optimism—the rate of forest loss peaked in the 1980s and has been declining since then.[84] Since 1500 we've lost around 0.5 percent to 1 percent of the world's vertebrate species; this is a rate of species loss that is much faster than the background rate of extinction and that meets or exceeds the rate during the earth's five mass extinction events.[85]

The trend in other non-wellbeing goods seems positive, however. We have made transformative scientific discoveries such as general relativity, quantum mechanics, and the theory of natural selection; we now know the age of the earth and the universe. And we have achieved some amazing things. Smallpox has been eradicated; we have climbed the highest mountains in the world, seen the tops of clouds from the vantage of powered flight, and photographed the earth from space. In 1900, 90 percent of the global population lived under autocratic rule; today more than half of people live in democracies.[86] Even the picture of global inequality is improving: although global inequality increased from 1800 until the 1970s, since then it has steeply declined thanks to rapid economic growth in Asia.[87] Because art is so subjective, it is nigh-on impossible to assess trends in artistic accomplishment, but one often-neglected factor is that, because of our sheer numbers, the artistic output of our species has increased dramatically: a higher population means more artists. And the artistic capacity of the population has, in some respects, greatly increased because of rising literacy and greater wealth: a more literate population has more writers, and the fewer people there are in dire poverty, the more artists there will be. In light of these considerations, it is likely that art has progressively reached new heights over time and will continue to do so at least for the next hundred years. The same applies for other non-wellbeing goods. The more people there are and the higher living standards are, the more likely it is that there will be individuals, like Usain Bolt, Margaret Atwood, or Maryam Mirzakhani, who go on to achieve great things.

How you evaluate these trends depends on the weight you put on non-wellbeing "bads," like destruction of the environment, and on "goods," like democracy and scientific progress. How to make this trade-off is a difficult question, and it's the sort of issue where it's hard for moral philosophy to provide illumination. My personal view is that the overall trend is positive.

The Case for Optimism

So far, I have examined whether the world has been getting better or worse over time. This has turned out to be fiendishly difficult. For all this, we now come to a harder task: to ascertain whether the world will get better or worse in the long run.

We can make some progress by focusing on just two extreme scenarios: the best or worst possible futures, *eutopia* and *anti-eutopia*. I call the worst possible world "anti-eutopia" because "dystopia" does not typically capture how bad the worst possible futures could be. For example, the dystopian scenarios that I envisaged in Chapter 4 and that are often discussed in science fiction would be bad, but they are optimised for things like the worship of the leader or the creation of a society in line with the leader's ideology, rather than to be *as bad as possible*.

Does considering these two possible futures give us grounds for optimism or pessimism? This depends on two things: the relative value of these worlds and how likely we are to realise them. The relative value of these worlds gives grounds for pessimism. In my view, the badness of the worst possible world is much greater than the goodness of the best possible world.

To make this intuitive, suppose you are faced with two options. The first is a gamble that gives you a 50 percent chance of creating the best possible eutopia for the future, with a huge civilisation consisting of the most flourishing possible lives, and a 50 percent chance of the worst possible anti-eutopia, with a huge civilisation consisting of lives suffering the most intense possible torment. The second option is to decline the gamble; if you do so, humanity will dwindle and then go extinct over the coming centuries. What would you do?

If the answer isn't clear, then consider just your own life. Imagine that you personally had the option of dying peacefully or a fifty-fifty chance of living in either eutopia, with the highest heights of flourishing, or anti-eutopia, with the deepest trenches of misery. I would certainly choose to die peacefully rather than to take the gamble, and I suspect that most people are the same.

It's not totally clear how to explain this intuition. Perhaps the intuitive asymmetry between happiness and suffering is due to nothing more than a fact of our biological makeup: as it happens, it is easier to produce pain than pleasure, so the worst experiences that we can possibly feel are much worse than the best experiences we can possibly feel. This asymmetry can potentially be explained on evolutionary grounds: from an evolutionary perspective, the downside of dying is much worse than the upside from eating a meal, say, or from a single act of sexual intercourse. So it would make sense that we would be far more strongly incentivised, through pain, to turn away from circumstances that might risk our death than we would be incentivised to turn towards a "good" like a meal or sex.

Perhaps, then, when we consider the best possible life or worst possible life, our imagination simply fails us: we just don't properly comprehend what the best possible life would be like. This gets some support from reflecting on peak experiences—the very best moments in life—and how we would trade such moments off against one another. That is, just what time span of experiencing the very worst moment would we accept in exchange for getting to experience the very best moments for a certain duration? For example, philosopher Bertrand Russell, in the prologue to his autobiography, wrote, "I have sought love . . . because it brings ecstasy—ecstasy so great that I would often have sacrificed all the rest of life for a few hours of this joy."[88] The Russian novelist Fyodor Dostoevsky described his experiences with epilepsy as follows:

> For several instants I experience a happiness that is impossible in an ordinary state, and of which other people have no conception. I feel full harmony in myself and in the whole world, and the feeling is so strong and sweet that for a few seconds of such bliss one could give up ten years of life, perhaps all of life.
>
> I felt that heaven descended to earth and swallowed me. I really attained god and was imbued with him. All of you healthy people don't even suspect what happiness is, that happiness that we epileptics experience for a second before an attack.[89]

If Dostoevsky is right, most people simply don't know how good life can be.

However, it might also be the case that the asymmetry is not just a product of our ignorance or our biology but is more deeply rooted in morality itself. Indeed, on a range of views in moral philosophy, we should weight one unit of pain more than one unit of pleasure. We already saw one possible route to this asymmetry in Chapter 8. I argued that when we are morally uncertain, we should adopt a critical level view, according to which a life needs to be sufficiently good for the person living it in order for the person's existence to make the world a better place. If this is correct, then in order to make the expected value of the future positive, the future not only needs to have more "goods" than "bads"; it needs to have *considerably* more goods than bads.

Overall, it seems to me we should think that the badness of the worst possible future is greater than the goodness of the best possible future. This brings us to the second question: How likely is it, relatively, that we will end up in eutopia rather than anti-eutopia? While my answer to the first question was pessimistic, I think there are grounds for optimism on the second.

The key argument for optimism about the future concerns an asymmetry in the motivation of future people—namely, people sometimes produce good things just because the things are good, but people rarely produce bad things just because they are bad. People often do things because they believe that these things are good for themselves, or good for others, or good for the world. So, for example, if someone spends their time travelling the world, or eating delicious food, or playing video games, we can explain this behaviour simply by noting that these things are good; similarly, if someone engages in social activism, we can explain this behaviour by noting that they believe it will make the world better.

In contrast, if we know that someone is undergoing a painful tooth operation, it's extraordinarily unlikely that they are doing this simply in order to have a bad time; rather, the bad experience is a necessary evil in order to avoid more pain later on. In general, even the worst atrocities typically have been committed not simply because they are bad but as a side effect of other actions or as a means to some other end. In an earlier section of this chapter, I described the suffering people currently inflict on nonhuman animals. People don't do this because they actively like the suffering of animals; rather, they like the taste of meat, want it cheaply, and aren't particularly concerned about the welfare of farmed animals, so they are willing to allow the suffering of animals to persist as a side effect. The same applies for other horrors that have been inflicted throughout history. Most people kept slaves not in order to make them suffer but in order to profit from their work, or as a status symbol. Wars are, in general, fought not in order to make the aggressor's opponents feel pain but to gain power and glory.

Sadly, this is not always true, and sadism has at times been widespread. Ordinary people thronged to see the gladiators in ancient Rome and to see gruesome public executions in early modern Europe. Moreover, some of the most influential figures in history have taken pleasure in the suffering of their victims.[90] Mao gave detailed instructions when ordering the torture and murder of millions of his victims, and he took pleasure in watching acts of torture.[91] Similarly, Hitler gave specific instructions for some of the plotters of the 1944 assassination attempt to be strangled with piano wire, and their agonizing deaths were filmed. According to Albert Speer, the minister of armaments and war production in Nazi Germany, "Hitler loved the film and had it shown over and over again."[92] But even in these cases, part of the motivation for these sadistic acts might have been to maintain power and signal status.

Although they are rare in the population as a whole, malevolent, sadistic, or psychopathic actors may be disproportionately likely to gain political power. Many dictators have exhibited such traits aside from Mao and Hitler, including Genghis Khan, Saddam Hussein, Stalin, Mussolini, Kim Il-sung, Kim Jong-il, François Duvalier, Nicolae Ceauşescu, Idi Amin, and Pol Pot. There is therefore a risk that malevolent people could have an outsize impact on the future.

Despite these important and worrying exceptions, in general people are much more often motivated to promote that which they believe to be good than that which they believe to be bad. We see this motivational asymmetry in current global expenditures.[93] Most spending is on the pursuit of things that are good: health, science, education, entertainment, and shelter. Only a small fraction of global expenditure is on imprisonment, war, factory farming, or other evils, and these are almost always done as a means to some other end.

This asymmetry in motivations is clear when we think about potential pathways to the best and worst possible futures. First, consider the best possible future: civilisation is full of beings with long, blissful, and flourishing lives, full of artistic and scientific accomplishment, expanded across the cosmos. We can come up with ready explanations of how such a civilisation might arise. A first explanation would invoke moral convergence: people in the future might have just recognised what is good and worked to promote the good. That is, over time, and with the enormous scientific and technological advances that the future might bring, including advances in the ability to reflect and reason with one another, everyone might have converged on a vision of what the best possible future is like and then put it into practice.

Second, even without moral convergence, people might have worked out their own visions of what a good life and good society consists of and co-operated and traded in order to build a society that is sufficiently good for everyone. The resulting society would be a compromise among different worldviews in which everyone gets most of what they want. Even if no one has a positive moral vision at all but just wants what's best *for them*, this could still result in a very good world. In a world where communication, trade, and compromise are easy and technology is extremely advanced, most people could get most of what they want.

Now, try to consider the worst possible civilisation: one that is as bad as the best possible future is good. Such a future would have to consist of an enormous number of people, spread out across the cosmos, living lives full of intense misery. Can we come up with explanations of how such an outcome could come about? It's much harder to do so. Realistic dystopian scenarios are usually optimised for some other end, not to make the world as bad as possible. So astronomically good futures seem eminently possible, whereas astronomically bad futures seem very unlikely.

The badness of anti-eutopia is greater than the goodness of eutopia, but eutopia is much more likely than anti-eutopia. All things considered, it seems to me that the greater likelihood of eutopia is the bigger consideration. This gives us some reason to think the expected value of the future is positive. We have grounds for hope.

PART V
TAKING ACTION

CHAPTER 10

What to Do

Backs to the Future

In the English language, the future is ahead of us and the past is behind. We might say that we must prepare for what lies before us and that we should not worry about what is behind us, or that we are facing a precarious future, or that Mary Wollstonecraft was a thinker ahead of her time. It turns out that this metaphorical mapping is near universal across cultures: as far as we know, every language in the world represents the future as being in front of us and the past as being behind, with just a handful of exceptions.[1]

The best-studied exception is the Aymara language. The Aymara are an Indigenous nation, comprising nearly two million people, who live in Bolivia, northern Chile, Argentina, and Peru.[2] Their traditional dress is brightly coloured, and their flag resembles technicolour glitch art. In the Aymara language, the future is behind us and the past is in front of us. So, for example, the phrase *nayra mara* is composed of the word for "front" (which also can refer to "eye" or "sight") and the word for "year," which means "last year." *Nayra pacha* literally means "front time" but refers to a "past time." To say "from now on," one says *akata qhiparu*, literally, "this from behind towards," and to refer to a "future day" one says *qhipüru*, literally, "behind day."

This conceptual metaphor is not restricted to Aymara speakers' choice of words. When referring to an event in the future, an Aymara speaker might point their thumb over their shoulder. This effect even persists when native Aymara speakers talk in a second language like Andean Spanish.

Almost all languages represent the future as ahead of us because when we walk or run, we both travel through time and travel forward through space. In the Aymara language, the more important feature of time is what we know and what we don't. We can *see* the present and the past; they are laid out before us. We can therefore have direct knowledge of them in a way we can't know the future—anything we know or believe about the future is based on inference from what we have experienced in the present or the past.[3] The implicit philosophy is that, when making plans for the future, we should take much the same attitude as if we were walking backwards into unknown terrain.

This metaphor is an appropriate way to think about our journey into the future. Over the last nine chapters, I hope I've shown that it's possible both to think clearly about the future and to help steer it in a better direction. But I'm not claiming it's easy. At best, I've given a quick over-the-shoulder glance at the future that lies behind us. There is still so much we don't know.

Even over the course of writing this book, I've changed my mind on a number of crucial issues. I take historical contingency, and especially the contingency of values, much more seriously than I did a few years ago. I'm far more worried about the longterm impacts of technological stagnation than I was even last year. Over time, I became reassured about civilisation's resilience in the face of major catastrophes and then disheartened by the possibility that we might deplete easily accessible fossil fuels in the future, which could make civilisational recovery more difficult.

We are often in a position of deep uncertainty with respect to the future for several reasons. First, for some issues, there are strong considerations on both sides, and I just don't know how they should be weighed against each other. This is true for many strategic issues around artificial intelligence. For example: Is it good or bad to accelerate AI development? On the one hand, slowing down AI development would give us more time to prepare for the development of artificial general intelligence. On the other hand, speeding it up could help reduce the risk of technological stagnation. On this issue, it's not merely that taking the wrong action could make your efforts futile. The wrong action could be disastrous.

The thorniness of these issues isn't helped by the considerable disagreement among experts. Recently, seventy-five researchers at leading organisations in AI safety and governance were asked, "Assuming that there will be an existential catastrophe as a result of AI, what do you think will be the cause?"[4] The respondents could give one of six answers: the first was a scenario in which a single AI system quickly takes over the world, as described in Nick Bostrom's *Superintelligence*; second and third were AI-takeover scenarios involving many AI systems that improve more gradually; the fourth was that AI would exacerbate risk from war; the fifth was that AI would be misused by people (as I described at length in Chapter 4); and the sixth was "other."

The typical respondent put a similar probability across the first five scenarios, with "other" being given a one-in-five chance. However, individual responses varied a lot, and the self-reported confidence in these estimates was low: the median respondent rated their own confidence level as a 2, on a scale from 0 to 6. There was even enormous disagreement about the size of the threat: when asked about the size of existential risk from AI, respondents gave answers all the way from 0.1 percent to 95 percent.[5]

Much the same is true of issues around AI governance. In 2021 Luke Muehlhauser, a grantmaker in AI governance at Open Philanthropy, commented, "In the past few years, I've spent hundreds of hours discussing possible high-value intermediate goals with other 'veterans' of the longtermist AI governance space. Thus I can say with some confidence that there is very little consensus on which intermediate goals are net-positive to pursue."[6]

The second reason why we face such deep uncertainty is that, as well as weighing competing considerations we're aware of, we also need to try to take into account the considerations we haven't yet thought of. In 2002, when talking about the lack of evidence of Iraqi weapons of mass destruction, US Secretary of Defense Donald Rumsfeld declared, "There are known knowns; there are things we know we know. We also know there are known unknowns; that is to say we know there are some things we do not know. But there are also unknown unknowns—the ones we don't know we don't know."[7]

Rumsfeld's comment was lampooned as obscurantism at the time, and it even earned a Foot in Mouth Award, which the Plain English Campaign bestows each year for "a baffling comment by a public figure."[8] But he was actually making an important philosophical point: we should bear in mind there may be considerations that we aren't even aware of.

To illustrate, suppose that a highly educated person in the year 1500 tried to make the longterm future go well. They would be aware of some relevant things, such as the persistence of laws, religions, and political institutions. But many issues wouldn't occur to them. The ideas that the earth's habitable life span could be a billion years and that the universe could be so utterly enormous, yet almost entirely uninhabited, would not have been on the table. Crucial conceptual tools for dealing with uncertainty, such as probability theory and expected value, had not yet been developed. They would not have been exposed to the arguments for a moral worldview in which the interests of all people are equal. They wouldn't have known what they didn't know.

The third reason why we face deep uncertainty is that, even in those cases where we know that a particular outcome is good to bring about, it can be very difficult to make that happen in a predictable way. Any particular action we take has a whole variety of consequences over time: some of these will be good, some will be bad, and many will be of unclear value. Nonetheless, ideally we should try to factor all the consequences we can into our decision.

When confronted with the empirical and evaluative complexity that faces us, it can be easy to feel clueless, as if there's nothing at all we can do. But that would be too pessimistic. Even if we're walking backwards into the future—and even if the terrain we're walking on is unexplored, it's dark and foggy, and we have few clues to guide us—nonetheless, some plans are smarter than others. We can employ three rules of thumb.

First, take actions that we can be comparatively confident are good. If we are exploring uncharted territory, we know that tinder and matches, a sharp knife, and first aid supplies will serve us well in a wide range of environments. Even if we have little idea what our expedition will involve, these things will be helpful.

Second, try to increase the number of options open to us. On an expedition, we would want to avoid getting stuck down a ravine we can't get out of, and if we weren't certain about the location of our destination, we would want to choose routes that leave open a larger number of possible paths. Third, try to learn more. Our expedition group could climb a hill in order to get a better view of the terrain or scout out different routes ahead.

These three lessons—take robustly good actions, build up options, and learn more—can help guide us in our attempts to positively influence the long term. First, some actions make the longterm future go better across a wide range of possible scenarios. For example, promoting innovation in clean technology helps keep fossil fuels in the ground, giving us a better chance of recovery after civilisational collapse; it lessens the impact of climate change; it furthers technological progress, reducing the risk of stagnation; and it has major near-term benefits too, reducing the enormous death toll from fossil fuel–based air pollution.

Second, some paths give us many more options than others. This is true on an individual level, where some career paths encourage much more flexible skills and credentials than others. Though I've been very lucky in my career, in general, a PhD in economics or statistics leaves open many more opportunities than a philosophy PhD. As I suggested in Chapter 4, keeping options open is important on a societal level, too. Maintaining a diversity of cultures and political systems leaves open more potential trajectories for civilisation; the same is true, to an even greater degree, for ensuring that civilisation doesn't end altogether.

Third, we can learn more. As individuals, we can develop a better understanding of the different causes that I've discussed in this book and build up knowledge about relevant aspects of the world. Currently there are few attempts to make predictions about political, technological, economic, and social matters more than a decade in advance, and almost no attempts look more than a hundred years ahead. As a civilisation, we can invest resources into doing better—building mirrors that enable us to see, however dimly, into the future that lies behind us.

Keeping these high-level lessons in mind, let's talk about what to do, starting with the question of which priorities to focus on.

Which Priorities Should You Focus On?

If you're on an expedition, there might be many problems facing you all at once: the tents leak; morale is low; a leopard is stalking you. You'd need to prioritise. The leaky tents might be annoying, but they're not as important as that leopard.

Similarly, when thinking about how to improve the world, the first step is to decide which problem to work on. When people are deciding how to do good, they often focus on a problem that is close to their heart, perhaps because someone they know is affected by it. Others focus on problems that are especially salient. But if our aim is to do as much good as possible, these intuitions may be a poor guide, because the highest-impact actions may be much more effective than typical actions.

To get a sense for which kind of things we're choosing between, let's first take stock of the threats I've mentioned in the previous chapters. First, the lock-in of bad values, perhaps precipitated by artificial general intelligence or the dominance of a single world ideology. Second, the end of civilisation, which could be brought about by war involving nuclear weapons or bioweapons, or made more likely by technological stagnation, depleting fossil fuel reserves, or greatly warming the planet. What can we do in each of these areas?

For some issues, we can take somewhat robustly good actions. This is true for climate change and fossil fuel depletion, where we can draw on huge amounts of relevant research on their physical basis, their socioeconomic effects, and policies for mitigation and adaptation. And, crucially, we have a yardstick we can use to compare different interventions. We know we are winning against climate change if carbon dioxide emissions decline, and the more the better. Each of us can encourage clean-tech innovation through political advocacy or by funding or working for effective nonprofits like Clean Air Task Force and TerraPraxis.

Biosecurity and pandemic preparedness is another area where we can do things that are robustly good, like promoting innovation to produce cheap and fast universal diagnostics and extremely reliable personal protective equipment. Organisations like the Johns Hopkins Center for Health Security and the Bipartisan Commission on Biodefense are helping to promote pandemic preparedness solutions internationally.

General disaster preparedness also seems robustly good. This can include things like increasing food stockpiles; building bunkers to protect more people from worst-case catastrophes; developing forms of food production not dependent on sunlight in case of nuclear winter; building seed vaults with heirloom seeds that could be used to restart agriculture;[9] and building information vaults with instructions for creating the technologies necessary to rebuild civilisation.

In other areas, the key priorities are to build up options and learn more. This is true of many issues around AI. We do not yet know what the AGI systems we're worried about are going to look like, except in their broad contours. This makes it hard to work on well-targeted solutions now, and because of the complex strategic situation, many well-intentioned attempts might even backfire.

The history of efforts to reduce AGI risk does illustrate, however, that there is at least one thing we can do in such a situation: building a field of morally motivated actors who can start reducing our uncertainty about what to do. Ten years ago, almost no one was working to positively steer the trajectory of AI. But there are now at least a hundred people working on this problem, and tens of millions of dollars are now spent on it every year.[10] Groups like the Center for Human-Compatible Artificial Intelligence and the Future of Humanity Institute have helped to build a field of researchers who are focused on safe AI development. The issue is also increasingly being taken seriously in technology policy, for instance by the Center for Security and Emerging Technology at Georgetown University in Washington, DC. This effort is still far too small, but it's growing.

The risk of great-power war is another example where field building and further research are key priorities. While there is a large body of work on the causes of war, we still have a lot to learn about practical ways to reduce the risks of war. For instance, we know that countries are more likely to go to war with each other if they have a long-standing rivalry or are geographic neighbours—especially if they have territorial disputes. But redrawing borders is hardly a feasible intervention, nor can we travel back in time to prevent countries from becoming rivals. And while we also know that democracies are less likely to fight each other, promoting democracy around the world is a major challenge. Given these uncertainties, identifying and training talented researchers and effective organizations who can improve our knowledge in this area strikes me as critical. Organisations like the Stockholm International Peace Research Institute may help us find the policies and programmes which, if implemented, give us the best chance at maintaining peace between great powers in the coming decades.

As well as improving our knowledge about particular issues, we can also try to get a better understanding of the implications of longtermism as a whole. For example, you can help find new crucial considerations. Perhaps there is an overlooked technology on the horizon that poses a grave threat to the survival of civilisation. Perhaps some changes to the world's institutions and cultures would be valuable trajectory changes. Either of these would be enormously important to identify. These and other crucial issues are worked on at places like the Global Priorities Institute, the Future of Humanity Institute, and Open Philanthropy.[11]

How should we choose which of these problems are most pressing? In Chapter 2, I suggested using the significance, persistence, and contingency framework to measure a problem's importance.

But we should not *only* consider a problem's importance: some problem might be very important even though there is very little that we can do about it. We can break this down into two components. First, *tractability*: How many resources would it take to solve a given fraction of the problem? Some problems are intrinsically easier to make progress on than others. For example, the use of chlorofluorocarbons (CFCs) posed an enormous problem to the world by depleting the ozone layer.[12] But the problem turned out to be comparatively easy to solve: there were a small number of companies that needed to get on board and good substitutes for CFCs.[13] It was fifteen years between scientists first discovering that CFCs could deplete the ozone layer and the Montreal Protocol, which phased out chlorofluorocarbons and essentially ended the problem.[14]

For climate change, the difficulty of international cooperation and the lack of good substitutes for fossil fuels make the problem much harder.[15] But at least the nature of the problem—burning fossil fuels releases carbon dioxide—is very clear. This means we can create metrics by which we can more easily track progress on the problem. For other areas, like moral progress or the safe development of artificial intelligence, things are murkier. The nature of the problem is disputed, and there aren't such clear metrics by which we can track success.

The second component is *neglectedness*. The greater the number of people working on a problem, the more likely it is that the low-hanging fruit—the best opportunities to do good—will be taken. If you work on more neglected problems, you can make a bigger difference.

For instance, philanthropists now spend billions of dollars on climate advocacy every year, governments and companies spend hundreds of billions addressing climate change, and it is one of the problems of choice for most young socially motivated people.[16] As I mentioned in Chapter 6, this is the main reason that the tide has started to turn on climate change. In contrast, issues around AI development are radically more neglected—though I noted that interest in the area is growing, philanthropic funding still amounts to only a few tens of millions of dollars a year, and there are only a couple of hundred people working in the area. This means that, if you can help make progress, you as an individual have the ability to be transformative in a way that is much harder in areas that have attracted more attention.

How to Act

Assuming that you have chosen the problem you think is most pressing, what do you do next? People often focus on personal behaviour or consumption decisions. The suggestion, implicit or explicit, is that if you care about animal welfare, the most important thing is to become vegetarian; if you care about climate change, the most important thing is to fly less and drive less; if you care about resource overuse, the most important thing is to recycle and stop using plastic bags.

By and large, I think that this emphasis, though understandable, is a major strategic blunder for those of us who want to make the world better. Often the focus on consumption decisions is accompanied by a failure to prioritise. Consider, for example, the recent wave of advocacy for reducing plastic. The total impact this has on the environment is tiny. You would have to reuse your plastic bag eight thousand times in order to cancel out the effect of one flight from London to New York.[17] And avoiding plastic has only a tiny effect on ocean plastic pollution. In rich countries with effective waste management, plastic waste very rarely ends up in the oceans. Almost all ocean plastic comes from fishing fleets and from poorer countries with less-effective waste management.[18]

Some personal consumption decisions have a much greater impact than reusing plastic bags. One that is close to my heart is vegetarianism. The first major autonomous moral decision I made was to become vegetarian, which I did at age eighteen, the day I left my parents' home. This was an important and meaningful decision to me, and I remain vegetarian to this day. But how impactful was it, compared to other things I could do? I did it in large part because of animal welfare, but let's just focus on its effect on climate change. By going vegetarian, you avert around 0.8 tonnes of carbon dioxide equivalent every year (a metric that combines the effect of different greenhouse gases).[19] This is a big deal: it is about one-tenth of my total carbon footprint.[20] Over the course of eighty years, I would avert around sixty-four tonnes of carbon dioxide equivalent.

But it turns out that other things you can do are radically more impactful. Suppose that an American earning the median US income were to donate 10 percent of that income, which would be around $3,000, to the Clean Air Task Force, an extremely cost-effective organisation that promotes innovation in neglected clean-energy technologies. According to the best estimate I know of, this donation would reduce the world's carbon dioxide emissions by an expected three thousand tonnes per year.[21] This is far bigger than the effect of going vegetarian for your entire life. (Note that the funding situation in climate change is changing fast, so when you read this, the Clean Air Task Force may already be fully funded. Giving What We Can keeps an up-to-date list of the best charities in climate and other areas.)

There are good reasons to become and stay vegetarian or vegan: doing so helps you be a better advocate for climate change mitigation and animal welfare, more able to avoid charges of hypocrisy; and you might reasonably think that avoiding causing unnecessary suffering is part of living a morally respectable life. But if your aim is to fight climate change as much as possible, becoming vegetarian or vegan is only a small part of the picture.

Emphasising personal consumption decisions over more systemic changes is often a convenient move for corporations. In 2019 Shell's chief executive, Ben van Beurden, gave a lecture in which he instructed people to eat seasonally and recycle more, lambasting people who eat strawberries in winter.[22] In reality, in order to solve climate change, what we actually need is for companies like Shell to go out of business. By donating to effective nonprofits, we can all make this kind of far-reaching political change much more likely.

Donations are more impactful than changing personal consumption decisions in other areas too. For example, in *Doing Good Better*, I argued that donating to the best global poverty charities is much more impactful than buying fair trade products. These examples are not a fluke. We should expect this pattern in almost all areas. The most powerful and yet simple reason is this: our consumption is not optimized for doing harm, and so by making different consumption choices we can avoid at most the modest amount of harm we'd be otherwise causing; by contrast, when donating we can choose whichever action *best* reduces the harm we care about. We can have as big an impact as possible by taking advantage of levers such as affecting policy.

Moreover, for many of the problems I have discussed in this book, it is just not possible to make any difference by changing your consumption behaviour. While each of us can mitigate climate change through our everyday actions, this is not true for the risk of a great-power war, engineered pandemics, or the development of AI. However, we can all work on these problems by donating to effective nonprofits. Whatever else you do in life, donations are one way to do an enormous amount of good.

Beyond donations, three other personal decisions seem particularly high impact to me: political activism, spreading good ideas, and having children.

The simplest form of political activism is voting. On the face of it, it is improbable that voting could really do a lot of good. Every election I have ever voted in would have turned out the same whether I had voted or not, and that is almost certainly true for everyone reading this book. What this line of reasoning neglects is that, even if the chance that you influence an election is small, the *expected* value can still be very high.[23] If you live in the United States in a competitive state, the chance that your vote will flip a national election falls between one in one million and one in ten million. As a rule of thumb, governments typically control around a third of a country's GDP. In the United States, the federal government spends $17.5 trillion every four years. The spending priorities of administrations overlap substantially, so your vote may influence perhaps only 10 percent of the budget. Even so, multiply the small probability of your vote making a difference in a national election with the enormous impact if your vote *does* make a difference, and your vote in a competitive state would influence an expected $175,000. And this is just considering the money you might affect. A bigger effect could come from harder-to-quantify factors such as the likelihood that different candidates will start a nuclear war. So even though the probability of flipping an election is small, the payoff can be big enough to make voting worthwhile.

There are several caveats to this. First, many voters do not live in competitive states. If you live in a state that's certain to go to a particular candidate, the expected value of voting might be tiny because the chance of your having an effect is so small. Second, to make your vote worthwhile, you need to do more than just turn up and vote; you need to be better informed and less biased than the median voter—otherwise you risk doing harm.

Many of the same arguments apply to other forms of political activism. Although the chance that you personally will make a difference by getting involved in a political campaign is small, the expected returns can be very high because if your campaign succeeds, the payoff could be very large.

Another way to improve the world is to talk to your friends and family about important ideas, like better values or issues around war, pandemics, or AI. This doesn't mean that you should promote these ideas aggressively or in a way that might alienate those you love. But discussion between friends has been shown to be one of the most effective ways to increase political participation,[24] and it is also probably a good way to get people motivated to work on some of the major problems of our time.

The final high-impact decision you can make is to consider having children. As I argued in Chapter 8, one mistake people sometimes make is to overemphasise the negative effects of having children and not to consider the benefits at all, both to the children and to the world. Although your offspring will produce carbon emissions, they will also do lots of good things, such as contributing to society, innovating, and advocating for political change. I think the risk of technological stagnation alone suffices to make the net longterm effect of having more children positive. On top of that, if you bring them up well, then they can be change makers who help create a better future. Ultimately, having children is a deeply personal decision that I won't be able to do full justice to here—but among the many considerations that may play a role, I think that an impartial concern for our future counts in favour, not against.

Career Choice

So far, I have looked at ways that you can use your time and money to improve the long term. But by far the most important decision you will make, in terms of your lifetime impact, is your choice of career. Especially among young people, it has become increasingly common to strive for positive impact as a core part of one's professional life rather than as a sideshow. More and more people don't just want money to pay their bills; they also want a sense of purpose and meaning.

This is why, as a graduate student, I cofounded 80,000 Hours with Benjamin Todd. We chose the name 80,000 Hours because that is roughly how many hours you have in your career: forty hours per week, fifty weeks per year, for forty years. Yet the amount of time that people normally spend thinking about their career is tiny in comparison. When that's combined with how poor existing career advice is, we end up with the outcome that a large proportion of people land in careers that are neither as fulfilling nor as impactful as they could be.

How, then, should you decide on a career? Again, we can return to our expedition metaphor. The three key lessons we identified were to learn more, build options, and take robustly good actions. These mirror the considerations that longtermists face when choosing a career:

1. **Learn:** Find low-cost ways to learn about and try out promising longer-term paths, until you feel ready to bet on one for a few years.
2. **Build options:** Take a bet on a longer-term path that could go really well (seeking upsides), usually by building the career capital that will most accelerate you in it. But in case it doesn't work out, have a backup plan to cap your downsides.
3. **Do good:** Use the career capital you've built to support the most effective solutions to the most pressing problems.

In reality, you'll be pursuing all of these priorities throughout your career, but each one will get different emphasis at different stages. Learning will tend to be most valuable early in your career. Building your options by investing in yourself and accruing career capital is most valuable in the early to middle stages of your career. Making a bet on how to do good is most valuable in the mid to late stages of your career. But your emphasis might move back and forth over time. For instance, a forty-year-old who decides to make a dramatic career change might go back into learning mode for a few years. And you might be lucky enough to find yourself with opportunities to have an enormous positive impact right out of college; if so, this framework shouldn't discourage you from doing that.

Let's first look at *learning*. People often feel a lot of pressure to figure out their best path right away. But this isn't possible. It's hard to predict where you'll have the best fit, especially over the long term, and if you're just starting out, you know very little about what jobs are like and what your strengths are. Moreover, even if you could find the best path now, it might change over time. The problems that are most pressing now could become less pressing in the future if they receive more attention, and new issues could be discovered. Likewise, you might find new opportunities to make progress that you hadn't anticipated.

Even your personal preferences are likely to change—probably more than you expect. Ask yourself, How much do you think your personality, values, and preferences will change over the next decade? Now ask, How much did they change over the previous decade? Intuitively, I thought they wouldn't change much over the next decade, but at the same time I think they changed a lot over the previous decade, which seems inconsistent. Surveys find similar results, which suggests that people tend to underestimate just how much they will change in the future.[25]

All of this means that it's valuable to view your career like an experiment—to imagine you are a scientist testing a hypothesis about how you can do the most good. In practical terms, you might follow these steps:

1. Research your options.
2. Make your best guess about the best longer-term path for you.
3. Try it for a couple of years.
4. Update your best guess.
5. Repeat.

Rather than feeling locked in to one career path, you would see it is an iterative process in which you figure out the role that is best for you and best for the world. The value of treating your career like an experiment can be really high: if you find a career that's twice as impactful as your current best guess, it would be worth spending up to half of your entire career searching for that path. Over time, it will become clearer whether you have found the right path for you. For many people, I think it would be reasonable to spend

5 percent to 15 percent of their career learning and exploring their options, which works out to two to six years.

Kelsey Piper provides one example of the value of learning early about your options. In order to test out her potential as a writer, while in college she wrote one thousand words a day for her blog.[26] It turned out that she was good at it. Blogging helped her figure out that writing was the right path for her and helped her to eventually get a job at Vox's Future Perfect, which covers topics relevant to effective altruism, including global poverty, animal welfare, and the longterm future.

When you are thinking about exploration, I think it is good to aim high, to focus on "upside options"—career outcomes that have perhaps only a one-in-ten chance of occurring but would be great if they did. Shooting for the moon is not always good advice. However, if you want to have a positive impact on the world, there's a strong case to be made for aiming high. Even if there is a small chance of success, the expected value of focusing on upside options can be great, and, crucially, there is a large skew in outcomes. In many fields, the most successful people are responsible for a large fraction of the impact; for example, various studies have found that the top 20 percent of contributors produce a third to a half of the total output.[27]

Even though focusing on upside options when you are exploring is very valuable, you should also limit the risk that you could do harm. Because we are so uncertain about longterm effects, there is an increased risk of doing harm, so you should take this consideration seriously. In a slogan: target upsides but limit downsides.

The next thing to consider on your career path is *building options* by investing in yourself. In a lot of fields, people's productivity peaks between ages forty and fifty.[28] So investing in career capital, in the skills and networks you need to have a big impact, is a top priority early in your career. Some of the skills you could focus on include the following:[29]

- Running organisations
- Using political and bureaucratic influence to change the priorities of an organisation
- Doing conceptual and empirical research on core longtermist topics

- Communicating (for example, you might be a great writer or podcast host)
- Building new projects from scratch
- Building community; bringing together people with different interests and goals

Investing in yourself can pay off in unanticipated ways. For example, based on 80,000 Hours's advice, Sophie decided not to apply to medical school and instead shifted her focus to global pandemics. She found funding for a master's degree in epidemiology to build career capital in the area. When COVID-19 broke out, she found a neglected solution: challenge trials, which can greatly speed up the development of vaccines by deliberately infecting healthy and willing volunteers with the novel coronavirus in order to test vaccine efficacy. So she co-founded 1DaySooner, a nonprofit that signed up thousands of volunteers for human challenge trials in order to speed up vaccine approval. The world's first challenge trial for COVID vaccines started in the UK in early 2021.[30]

There is sometimes a trade-off between exploring and investing. This is particularly clear in academia. If I wanted to try out a different job and quit academic philosophy for a few years, that would probably be the end of my philosophy career—in my field, once you leave there is no way back. But things are not usually as clear-cut as this, and building career capital does not always preclude exploring later on.

The final consideration for choosing a career is the one we ultimately care about: *doing good*. For most people, the opportunity to have a lot of impact comes later in their career, once they have gained career capital. But sometimes you might come across a great opportunity to do good right away. For instance, Kuhan Jeyapragasan realised that his position as a student at Stanford University gave him a great platform for spreading awareness of important ideas. He helped to start the Stanford Existential Risk Initiative, which has helped hundreds of people learn about risks to humanity's longterm future.

In large part, how much good you do depends on the problem you choose to work on. As I argued earlier, there are probably very large differences in impact between problem areas, so making this choice carefully is crucial. The immediate impact you have will also be determined by the quality of the project you are working on, your seniority, and the strength of your team.

The "learn more, build options, do good" framework is generally useful for anyone deciding what to do with their career. But the specific path that works best for *you* depends on your *personal fit*. Some people are happiest locked away for months on end researching abstruse topics in economics or computer science, while others excel at managing a team or communicating ideas in a simple and engaging way.

You might also have some unique opportunities that other people don't have. Marcus Daniell is a professional tennis player from New Zealand. He is one of the top fifty doubles players in the world, and he won a bronze medal in doubles at the 2021 Tokyo Olympics. After learning about effective altruism, Marcus set up High Impact Athletes, which encourages professional athletes to donate to effective charities working on global development, animal welfare, and climate change. People who have donated through High Impact Athletes include Stefanos Tsitsipas, the current number four tennis player in the world, and Joseph Parker, a former world heavyweight champion boxer and sparring partner for Tyson Fury. The opportunity to set up High Impact Athletes was unique to Marcus; his network allowed him to try out something new and set up an organisation with lots of potential upside.

Isabelle Boemeke's story is in some ways similar. She started out as a fashion model, but after speaking to experts who said nuclear energy was needed to tackle climate change but were afraid to promote it because of its unpopularity, she pivoted to using her social media following to advocate for nuclear power. Of course, I'm not recommending professional tennis or fashion modelling as reliably high-impact careers, but these examples illustrate the importance of focusing on where you personally, with all your unique skills and abilities, can make the biggest difference on the world's most pressing problems. It would, for instance, have made little sense for Marcus or Isabelle to retrain as an epidemiologist or a climate scientist.

For many people, personal fit can mean the best way of contributing is through donations: you work in a career you love and excel at, and even if the work itself is not hugely impactful, you can make an enormous difference with your giving. This was true of John Yan. After learning about effective altruism and thinking about his career options, he decided to continue as a software engineer and donate a significant fraction of his income to effective charities as a member of Giving What We Can.[31]

Personal fit is a crucial determinant of your career's impact—it is a force multiplier on the direct impact you have and on the career capital that you gain. As mentioned before, outcomes are heavily skewed. If you can be in the top 10 percent of performers in a role rather than in the top 50 percent, this could have a disproportionate effect on your output. Being particularly successful in a role also gives you more connections, credentials, and credibility, increasing your career capital and leverage.

Personal fit is, in addition, one of the main ingredients of job satisfaction. People often associate altruism with self-sacrifice, but I think that for the most part, that is the wrong way to think about it. For me personally, since I started trying to do the most good with my life, I feel that my life is more meaningful, authentic, and autonomous. I am part of a growing community of people trying to make the world a better place, and many of these people are now among my closest friends. Effective altruism has added to my life, not subtracted from it. There is, moreover, a pragmatic reason to do a job you enjoy: it makes your impact sustainable over the long term. You want to be able to sustain your commitment to doing good for over forty years rather than think about how you can do as much good as possible this year. The risk of burnout is real, and you will work better with other people and be more productive if you are not stressed or depressed.

Doing Good Collectively

I've argued that positively influencing the longterm future is a key moral priority of our time. But it's not the only thing that matters. We should try to make the longterm future better in the context of living a rounded ethical life.

As part of this, it's particularly important to avoid doing harm. History is littered with people doing bad things while believing they were doing good, and we should do our utmost to avoid being one of them. For example, consider the Animal Rights Militia, which in the 1980s and '90s in the UK sent letter bombs to members of Parliament, including the prime minister at the time, and used bombs to set fire to buildings across the UK. Those behind these actions presumably thought they were acting morally—doing what was needed to reduce the suffering of animals. But they were wrong, and not just in this instance: doing significant harm to serve the greater good is very rarely justified. Here is why.

First, naive calculations that justify some harmful action because it has good consequences are, in practice, almost never correct. The Animal Rights Militia might have thought they were doing what was best for animals, but in reality they were hindering the cause by tainting it with violent extremism. This is particularly true when we consider that there are often a wide variety of ways of achieving a goal, many of which do not involve doing harm. The best alternative for the Animal Rights Militia wasn't sitting at home and doing nothing: it was engaging in peaceful and nonviolent protest and campaigning.

Second, plausibly it's wrong to do harm even when doing so will bring about the best outcome. This is an issue that divides what are called "consequentialists" and "nonconsequentialists" in moral philosophy. Even if you are sympathetic to consequentialism—in which the ends are all that ultimately matter—given the difficulty of ethics, you should not be certain in that view. And when we are morally uncertain, we should act in a way that serves as a best compromise between different moral views.[32] If one reasonable view says that avoiding harm is very important, then we should put significant weight on that when we act.

Similar considerations apply to other commonsense moral considerations. You might reason in a particular case that lying would produce the best consequences, but lying has many indirect negative effects that are difficult to observe, and it's plausibly intrinsically wrong too. So, in practice, I think it makes sense to almost never lie, even when it seems like doing so would be for the best. For similar reasons, one should strive to be a good friend and family member and citizen, to act kindly, and to cultivate a habit of cooperation—even if, in any given situation, it is not clear why this would lead to the best possible outcome. In these ways, I see longtermism as a supplement to commonsense morality, not a replacement for it.

A different way in which naive expected-value reasoning can lead us astray is if we think too individualistically, paying attention only to what we as individuals can achieve rather than thinking in terms of what the whole community of people engaged in longtermism can do.

I have seen the importance of group action firsthand through the effective altruism community. Since it was formed a decade ago, this community has grown to thousands of members who share information and opportunities, have their own online forum to discuss the latest ideas, and provide friendship and social support for one another. Undoubtedly, the community is more than the sum of its parts: we can achieve far more by working together than we would if we each tried to do good on our own. Importantly, because this community has a shared aim of doing the most good, I have reasons to help others in the community even if I do not receive anything in return.

The fact that we each act as part of a wider community warrants a "portfolio approach" to doing good—taking the perspective of how the community as a whole can maximize its impact. Then you can ask what you can do to move the community closer to an ideal allocation of resources, given everyone's personal fit and comparative advantage. Taking a community perspective, the primary question becomes not "How can I personally have the biggest impact?" but "Who in the community is relatively best placed to do what?" For example, my colleague Greg Lewis believes that AI risk is the most important issue of our time. But he thinks the risk from engineered pandemics is also important, and because he has a medical degree, it makes more sense for him to focus on that threat and let others focus on AI.

The portfolio approach can also give greater value to experimentation and learning. If one person pursues an unexplored path to impact (such as an unusual career choice), everyone else in the community gets to learn whether that path was successful or not. It can also give much greater value to specialisation: a community of three people might need only generalists, but a community of thousands will need people with particular specialist skills.

The portfolio approach also makes it easier to see how you can have an impact. If you only consider what you personally might be able to achieve, it is easy to feel powerless in the face of huge international problems like climate change and engineered pathogens. But if you instead ask "Would we make progress on the threat from engineered pandemics if there were hundreds of motivated and smart people working on it?" I think it becomes clear that the answer is yes.

Building a Movement

This chapter has discussed many ways you can directly have impact. But you can also go "meta": spread the idea of longtermism itself and convince others to care about future generations, to take the scale of the future seriously, and to act to positively influence the long term. You can do this by writing, organizing, talking to people you know, or getting involved with organisations such as 80,000 Hours and the Centre for Effective Altruism, where movement building is a component of their work.

Spreading these ideas can be an enormously powerful way of having an impact. Suppose that you convince just one other person to do as much good as you otherwise would have done in your life. Well, then you've done your life's work. Convince two other people, and you've tripled your impact.

Of course, we can take this reasoning too far. There are limits to how big a longtermist movement could be. And ultimately, movement building isn't enough: we need to actually solve the problems I've discussed.

But the nascency of longtermism suggests that developing and spreading ideas around it should be a core part of the movement's portfolio. For many previous social movements, change took time. The first public denouncement of slavery by the Quakers—the Germantown petition—was in 1688.[33] The Slavery Abolition Act in the British Empire was passed only in 1833, and several countries abolished slavery after 1960. Success took hundreds of years.

So, too, with feminism. Mary Wollstonecraft is often regarded as the first English-language feminist.[34] Her seminal work, *A Vindication of the Rights of Woman*, was published in 1792. The United States and the UK only gave men and women equal voting rights in 1920 and 1928, respectively, and it was only in 1971 that Switzerland did the same.[35] And of course, there is still much further to go on women's rights.

We may not see longtermism's biggest impacts in our lifetimes. But by advocating for longtermism, we can pass the baton to those who will succeed us—those who might run faster, see farther, and achieve more than we ever could. They will have the benefits of decades' more thought on these issues. And perhaps crucial moments of plasticity, when the direction of civilisation will be set, will occur during their lives rather than ours.

Recent history should give us hope that the world will start taking the interests of future generations seriously. Environmentalists have made the wellbeing of future generations salient in a way that has had real impact. To take just one example: After decades of campaigning, in 1998 the Greens became part of the coalition government in Germany, and in 2000, they introduced landmark legislation that would almost singlehandedly underwrite the global solar industry's growth, making Germany the world's largest solar market. By 2010, Germany accounted for nearly half of the global market for solar deployment.[36] From the perspective of providing power to Germany alone—a northern-latitude and fairly cloudy country—this made little sense. But from a global perspective, it was transformative. Thanks to this and other subsidy schemes introduced around the same time, the cost of solar panels fell by 92 percent between 2000 and 2020.[37] The solar revolution that we're about to see is thanks in large part to German environmental activism.[38]

I've seen successes from those motivated explicitly by longtermist reasoning, too. I've seen the idea of "AI safety"—ensuring that AI does not result in catastrophe even after AI systems far surpass us in the ability to plan, reason, and act (see Chapter 4)—go from the fringiest of fringe concerns to a respectable area of research within machine learning. I've read the UN secretary-general's 2021 report, *Our Common Agenda*, which, informed by researchers at longtermist organisations, calls for "solidarity between peoples and future generations."[39] Because of 80,000 Hours, I've seen thousands of people around the world shift their careers towards paths they believe will do more longterm good.

But we should not be complacent. There are enormous challenges ahead. We need to decarbonise the economy over the next fifty years, even as energy demand triples.[40] We need to reduce the risks of war between great powers, of the use of engineered pathogens, and of AI-assisted perpetual global totalitarianism. And at the same time, we need to ensure that the engine of technological progress keeps running.

If we are to meet these challenges and ensure that civilisation at the end of this century is pointed in a positive direction, then a movement of morally motivated people, concerned about the whole scope of the future, is a necessity, not an optional extra.

Who should this movement consist of? Well—if not you, then who?[41]

Positive moral change is not inevitable. It's the result of long, hard work by generations of thinkers and activists. And if there's any change that's not inevitable, it's concern for future people—people who, by virtue of their location in time, are utterly disenfranchised in the world today.

If we are careful and far-sighted, we have the power to help build a better future for our great-grandchildren, and their great-grandchildren—down through hundreds of generations. But we cannot take such a future for granted. There's no inevitable arc of progress. No *deus ex machina* will prevent civilisation from stumbling into dystopia or oblivion. It's on us. And we are not destined to succeed.

Yet success is possible—at least if people like you rise to the challenge. You may have more power than you realise. If your income is more than $20,000 per year (post-tax, with no dependents), then you are in the richest 5 percent of the world's population, even after adjusting for the fact that money goes further in lower-income countries.[42] And you probably live in one of the more powerful countries in the world, where you can campaign to change the attitudes of your conationals and the policies of your government.

If you've read this far, then probably you *care*, too. The last ten chapters have not been easy. Since you've made it through discussions of impossibility theorems in population ethics and of weighing chicken suffering against human happiness, you probably were convinced enough by my arguments in the first chapters that you wanted to know how it would all pan out—what the practical upshot would be. If there's ever anyone who will take action on behalf of future generations, it's you.

But can one person make a difference? Yes. Mountains erode because of individual raindrops. Hurricanes are just the collective movement of many tiny atoms. Abolitionism, feminism, and environmentalism were all "merely" the aggregate of individual actions. The same will be true for longtermism.

We've met some people who made a difference in this book: abolitionists, feminists, and environmentalists; writers, politicians, and scientists. Looking back on them as figures from "history," they can seem different from you and me. But they weren't different: they were everyday people, with their own problems and limitations, who nevertheless decided to try to shape the history they were a part of, and who sometimes succeeded. You can do this, too.

Because if not you, who? And if not now, when?

Out of the hundreds of thousands of years in humanity's past and the potentially billions of years in her future, we find ourselves living *now*, at a time of extraordinary change. A time marked by the shadow of Hiroshima and Nagasaki, with thousands of nuclear warheads standing ready to fire. A time when we are burning through our finite fossil fuel reserves, producing pollution that might last hundreds of thousands of years. A time when we can see catastrophes on the horizon—from engineered pathogens to value lock-in to technological stagnation—and can act to prevent them.

This is a time when we can be pivotal in steering the future onto a better trajectory. There's no better time for a movement that will stand up, not just for our generation or even our children's generation, but for all those who are yet to come.

Acknowledgements

I could not possibly have written this book alone. Literally hundreds of people helped shape the words on these pages. I am grateful for the advice, knowledge, feedback, and inspiration they provided.

I'm extraordinarily grateful to have a team of talented, committed people work with me; I'm humbled that I get a chance to work with people who inspire me every day. Laura Pomarius, Luisa Rodriguez, and Max Daniel each (at different times) worked as my chief of staff, leading the team that worked on the book and managing the whole project. Frankie Andersen-Wood and Eirin Evjen worked (at different times) as my executive assistant, providing invaluable support to me and others on the team. Aron Vallinder, John Halstead, Stephen Clare, and Leopold Aschenbrenner were research fellows, doing much of the research underlying the book. The manifold ways in which cach of these team members have improved the book are almost impossible to compute; it would never have happened without them.

Some people were not part of the core team but acted as regular advisers. Joe Carlsmith improved the language greatly in many sections and provided insightful advice on many of the key decisions governing the book. A. J. Jacobs provided advice on writing style and storytelling, and conducted some interviews. Anton Howes provided general guidance on history and first alerted me to the abolition of slavery as a significant, persistent, and contingent historical event. Peter Watson and Danny Bressler advised me on climate change. Christopher Leslie Brown guided me through the scholarship on abolition from the early stages of my work. Ben Garfinkel advised me on AI. Lewis Dartnell advised on collapse and recovery. Carl Shulman advised on many issues, including stagnation and collapse and recovery.

The research assistance I got from my team and advisers was very substantial, and many sections of the book were essentially coauthored. These sections are: Chapter 1, "Future People Count" (with Joe Carlsmith); Chapter 3, "The Contingency of Values" (with Stephen Clare); Chapter 4, "The Hundred Schools of Thought" (with Tyler John) and "How Long Till AGI?" (with Max Daniel); Chapter 5, "Spaceguard" (with John Halstead) and "Great-Power War" (with Stephen Clare); Chapter 6, "The Historical Resilience of Global Civilisation," "Would We Recover from Extreme Catastrophes?" (both with Luisa Rodriguez), "Climate Change," and "Fossil Fuel Depletion" (both with John Halstead); Chapter 7 on stagnation in its entirety (with Leopold Aschenbrenner); Chapter 9, "Are People Getting Happier?" and "Non-wellbeing Goods" (both with John Halstead); Chapter 10, "How to Act" and "Career Choice" (both with John Halstead); and the appendix, "The SPC Framework" (with Teruji Thomas and Max Daniel).

Almost all the graphs and figures in this book were created by Taylor Jones and Fin Moorhouse, who did a terrific job at tracking down data, visualising them in a way tailored to this book, and being patient when responding to my team's repeated requests for revisions.

The immense task of fact-checking each and every sentence was done by Joao Fabiano, Anton Howes, Max Daniel, Stephen Clare, and John Halstead. Most of the sixty thousand words of endnotes were written by Max, Stephen, and John. Joao also did the bibliography and reference database.

I also want to acknowledge major contributions to parts of the book that never made it into the final manuscript. This includes Tyler John on longtermist institutional reform, Jaime Sevilla on persistence studies, and Luisa Rodriguez on forecasting. And I want to acknowledge those who had a particularly profound impact on my broader thinking about longtermism, especially Toby Ord, Holden Karnofsky, Carl Shulman, and Hilary Greaves. Their influence on me is so thoroughgoing that it permeates every chapter.

I got particularly helpful advice on writing style and structure from Brian Christian, Dylan Matthews, Jim Davies, Larissa MacFarquhar, Rutger Bregman, and Max Roser.

I have also benefited immensely from thoughtful and detailed comments on the book from dozens of expert reviewers. This book draws on everything from paleoclimatology to the history of Confucianism. I could not have hoped to do this range of topics justice without feedback and advice from topic experts: Dr Leslie Abrahams (climate change), Dr Wladimir Alonso (animal welfare), Prof. Eamon Aloyo (great-power war), Prof. Jutta Bolt (economic history), Prof. Robert Boyd (cultural evolution), Prof. Bear Braumoeller (great-power war), Prof. Christopher Brown (history of slavery), Dr Sally Brown (climate change), Prof. Matthew Burgess (climate change), Prof. Paul Burke (climate change), Prof. Bryan Caplan (population ethics), Dr Lucius Caviola (psychology of wellbeing), Dr Paulo Ceppi (climate science), Prof. David Christian (history), Prof. Antonio Ciccone (climate change), Prof. Matthew S. Clancy (economic stagnation), Dr Paul Collins (Sumerian Empire), Prof. Tyler Cowen (economic stagnation), Dr Colin Cunliff (public policy), Dr Allan Dafoe (great-power war), Prof. Lewis Dartnell (civilisational collapse and recovery), Prof. Hadi Dowlatabadi (climate change), Dr David Edmonds (population ethics), Prof. Kevin Esvelt (biosecurity), Grethe Helene Evjen (civilisational collapse), Prof. Laura Fortunato (cultural evolution), Derek Foster (subjective wellbeing), Prof. Chris Fraser (Mohism), Dr Goodwin Gibbins (climate science), Prof. Colin Goldblatt (climate change), Prof. Paul Goldin (Chinese history), Solomon Goldstein-Rose (climate change), Prof. Donald Grayson (megafauna), Prof. Joshua Greene (psychology of wellbeing), Prof. Johan Gustafsson (population ethics), Dr Jonathan Haas (hunter-gatherers), Prof. Joanna Haigh (climate change), Prof. Kenneth Harl (Roman Empire), Prof. Alan Harris (asteroids), Prof. David Hart (public policy), Dr Zeke Hausfather (climate change), Prof. Gary Haynes (megafauna), Prof. Cecilia Heyes (cultural evolution), Ziya Huang (China), Dr Matthew Ives (climate change), Prof. Mattias Jakobsson (genetics), Dr Kyle Johanssen (animal welfare), Dr Toshiko Kaneda (demography), Prof. J. Paul Kelleher (climate change), Prof. Morgan Kelly (persistence studies), Prof. Robert Kelly (hunter-gatherers), Dr Matt Killingsworth (psychology of wellbeing), Prof. Pamela Kyle Crossley (Chinese history), Dr Jerome Lewis (hunter-gatherers), Prof. Emily Lindsey (megafauna extinction), Prof. Marc Lipsitch (epidemiology), Prof. Marlize Lombard (archaeology), Prof. Jonathan Losos (evolution of life on Earth and the Fermi paradox), Prof. Heike Lotze (climate change), Prof. Dan Lunt (climate science), Prof. Kathleen Lyons (megafauna), Prof. Andrew MacDougall (climate science), Dr David Mathers (animal welfare), Dr Linus Mattauch (climate economics), Prof. Jeff McMahan (Derek Parfit), Prof. David Meltzer (megafauna), Prof. Alex

Mesoudi (cultural evolution), Prof. Ron Milo (environmental sciences), Dr Kieren Mitchell (megafauna), Dr Steve Mohr (climate change), Dr Dimila Mothé (megafauna extinction), Prof. Dani Nedal (great-power war), Prof. Robert Nicholls (climate change), Dr Tessa Peasgood (subjective wellbeing), Dr Angela Perri (megafauna extinction), Prof. Osvaldo Pessoa (philosophy of science), Dr Max Popp (climate change), Prof. Dudley Poston (demography), Prof. Rachell Powell (evolution of life on Earth and the Fermi paradox), Prof. Imants Priede (zoology), Prof. Ramses Ramirez (climate change), Dr Colin Raymond (climate science), Dr Justin Ritchie (climate change), Prof. Tapio Schneider (climate change), Dr Cynthia Schuck-Paim (animal welfare), Dr Oliver Scott Curry (anthropology), Prof. Jeff Sebo (animal welfare), Dr Mikhail Semenov (climate change), Dr Rohin Shah (AI), Prof. Steven Sherwood (climate science), Dr Adam Shriver (animal welfare), Dr Peter Spreeuwenberg (public health), Prof. Amia Srinivasan (Derek Parfit), Prof. Chris Stringer (anthropology), Dr Jessie Sun (psychology of wellbeing), Ted Suzman (political advocacy), Prof. Michael Taylor (history of slavery), Prof. William Thompson (Sumerian Empire), Philip Thomson (hunter-gatherers), Prof. Bryan Ward-Perkins (Roman history), Prof. Andrew Watson (climate change), Dr Peter Watson (climate science), Dr Mark Webb (climate change), Dr Daniel Welsby (climate change), Prof. Paul Wignall (megafauna), Prof. Greg Woolf (Roman Empire), and the World Energy Outlook Team (climate change). Thank you all. These advisers don't necessarily agree with the claims I make in the book, and all errors in the book are my responsibility alone.

Many others also took the time to read and give insightful and detailed feedback on earlier drafts of this book. I am grateful for helpful comments from Abie Rohrig, Alejandra Padin-Dujon, Alex Moog, Alexander Berger, Alimi Salifou, Allen Dafoe, Allison Wilkinson, Ana Gonzalez Guerrero, Andreas Mogensen, Andrew Alonso y Fernandez, Andrew Leigh, Angela Aristizabal, Angus Mercer, Ann Garth, Anna Mohan, Arden Koehler, Arthur Wolstenholme, Arushi Gupta, Astrid Olling, Asya Bergal, Becca Segal, Ben Garfinkel, Ben Hoskin, Ben Todd, Benjamin Glanz, Benny Smith, Brian Tomasik, Brian Tse, Caleb Parikh, Cameron Mayer Shorb, Carl Shulman, Cate Hall, Christian Tarsney, Cindy Gao, Clíodhna Ní Ghuidhir, Cullen O'Keefe, Damon Binder, Danny Bressler, Dave Bernard, David

Manheim, David Roodman, Douglas Rogers, Elise Bohan, Eric Sorge, Eva Vivalt, Fin Moorhouse, Garrison Lovely, Greg Lewis, Gully Bujak, Habiba Islam, Hamish Hobbs, Hannah Bartunik, Hannah Ritchie, Hannah Wang, Harri Besceli, Hayden Wilkinson, Heather Marie Vitale, Helen Toner, Holden Karnofsky, Iain Crouch, Isaac Dunn, Isabel Juniewicz, Jacob Barrett, Jacob Eliosoff, Jade Leung, Jakob Sønstebø, Jamie Harris, Jason Crawford, Jeff Alstott, Jennifer Mack, Jess Whittlestone, Jesse Clifton, Johannes Ackva, Josef Nasr, Joseph Carlsmith, Joshua Monrad, Julia Wise, Kaleem Ahmid, Katie Lyon, Keirra Woodward, Kimya Ness, Kirsten Horton, Koji Flynn Do, Kuhan Jeyapragasan, Laura Pomarius, Lexi Caruthers, Linh Chi Nguyen, Linda Doyle, Lizka Vaintrob, Lucius Caviola, Luisa Sandkühler, Luke Muehlhauser, Malo Bourgon, Mark Devries, Matthew van der Merwe, Max Roser, Max Xu, Medhavi Gupta, Michelle Hutchinson, Mike Levine, Moritz Adam, Naomi Pyburn, Natalie Cargill, Nick Beckstead, Nicole Ross, Ollie Base, Orlando van der Pant, Owen Cotton-Barratt, Pablo Stafforini, Paul Christiano, Pernille Brams, Philipp Trammell, Richard Ngo, Rob Long, Robin Lintz, Rohin Shah, Rūta Karolytė, Sabrina Baier, Sashika Coxhead, Shankar Charithran, Shreedhar Manek, Sohum Pal, Stefan Schubert, Stefan Torges, Sumaya Nur, Tena Thau, Toby Newberry, Toby Ord, Tom Critchley, Tom Davidson, Tom Moynihan, Tyler John, Victor Warlop, Vishwa Prakash, Xuan, Zachary Brown, and Zarah Baur.

I thank Oxford University, the Faculty of Philosophy, and the Global Priorities Institute for providing such a wonderful institutional home and giving me the flexibility I needed to undertake such a major project.

I owe an enormous debt to Cecilia Stein for being an extraordinary editor—for providing so many rounds of in-depth commentary, for pushing me to find the personal in the abstract, and for being such a champion of this book. Thank you also to Alex Christofi, to the rest of the Oneworld team, and to TJ Kelleher, Jessica Breen, Stewart Hendricks, Jenny Lee, and the rest of the team at Hachette, for backing this project. Thank you to my agent, Max Brockman, for helpful advice throughout the process.

I send my love and thanks to my parents, Mair and Robin, and my brothers, Iain and Tom. I've always felt supported and loved by you. And I send the same love and thanks to the many friends who've made the times when I've not been writing so fun, including Amanda, Chris, Cleo, George,

Georgie, Kev, Matthieu, Rinad, Robbie, and Siobhan. And thank you, Elif, for cultivating my love of music and India and Knockout, for helping me be a human, and for your encouragement as I first started this work.

During the course of writing this book I lived in five different houses, and my housemates (and house-bubbles) were an endless source of joy. Thank you, Simeon, for being willing to sacrifice yourself to those bulls in order to save my life. Thank you, Natalie, for always backing me, mostly refraining from violence against me, and convincing me to keep the Easter egg. Thank you, Liv and Igor, for ending my uncompetitive streak, for unfettered prances round the garden, and for keeping life absurd. Thank you, Laura and Luisa, for holding a potato-themed party in my honour. Thank you, Hamish and Anthony, for endless games of bananarchy, Miranda surprises, and the filthy drop. Thank you, Rūta and Elly, for the skanking.

Finally, thank you, Holly, for being so constructively caring and so uncompromisingly supportive. By all rights, the two years of the pandemic should have been miserable. They were the best of my life. Thank you for giving me a slice of eutopia; a taste of just how good life can be.

Appendices

1. Further Resources

The book's website is at whatweowethefuture.com. It includes supplementary materials and an up-to-date list of further reading.

For career advice and a podcast featuring unusually in-depth conversations about the world's most pressing problems, see 80000hours.org.

If you want to take a pledge to donate to charity, go to www.givingwhatwecan.org.

For more information about longtermism, see longtermism.com. For more information about effective altruism, see effectivealtruism.org.

For a window into the thinking of two of the people who have most influenced my views on longtermism, see Toby Ord's (2020) *The Precipice* and Holden Karnofsky's blog *Cold Takes* (cold-takes.com).

2. Terminology

This book defends and explores the implications of *longtermism*, the view that positively influencing the longterm future is one of the key moral priorities of our time. It should be distinguished from *strong longtermism*, the view that positively influencing the longterm future is *the* moral priority of our time—more important, right now, than anything else.

I explore the case for strong longtermism in an academic article with my colleague Hilary Greaves.[1] The case is surprisingly strong, given how neglected longterm issues currently are, but it's sensitive to a number of very tricky philosophical issues, such as how to take into account very small probabilities, how to act in the face of highly ambiguous evidence, and how much sacrifice is required from the present generation for the sake of future generations.[2] It's not a view we should be highly confident in, and I don't defend it in this book.

I suggest that there are two ways of positively influencing the longterm future: first, by effecting positive *trajectory changes*, which increase the average value of future civilisation over its life span, improving future civilisation's "quality of life"; and second, by *ensuring civilisation's survival*, increasing its life span.

An alternative framing is given by the idea of existential risks, which are "risks that threaten the destruction of humanity's longterm potential."[3] This concept is important and useful in many contexts. But I tend not to use it because much of my focus is on improving the values that guide the future, and for two reasons this idea doesn't fit neatly under the category of existential risk reduction. First, by improving future values, one can make the future better, but this does not involve preventing the "destruction" of humanity's longterm potential; the improvement to future values might only be small. Second, if bad values guide future civilisation, humanity can retain its "potential" (because future leaders *could* adopt better values, if they chose to) while losing out on almost all actual value (because those leaders *do not* choose to adopt better values). But it's what actually happens that we should care about, not what future people have the potential to make happen.

3. The SPC Framework

In the book I give a framework for assessing the longterm value of bringing about a state of affairs, which I state as follows:

Significance is the average value of that state of affairs over time.

Persistence is how long that state of affairs lasts.

Contingency is the proportion of that time that the world would not have been in this state of affairs anyway.

We can define this formally. Consider some possible action aimed at bringing about some state of affairs s. Let p be the effect of that action and let q be the status quo—what would happen if we took no action.[4] $V_s(p)$ is the total value contributed from being in state *s*, given *p*; $V_s(q)$ is the total value contributed from being in state *s*, given *q*. $T_s(p)$ is the length of time that the world is in state *s*, given *p*; $T_s(q)$ is the length of time that the world is in state *s*, given *q*.

Significance $=_{df} [V_s(p) - V_s(q)] / [T_s(p) - T_s(q)]$

Persistence $=_{df} T_s(p)$

Contingency $=_{df} [T_s(p) - T_s(q)] / T_s(p)$

These three terms multiply together to give $V_s(p) - V_s(q)$, or the total value contributed from being in a state of affairs *s*, given *p* rather than *q*. That is: significance × persistence × contingency = longterm value.

Because these multiply, we can intuitively compare different longterm effects: between two alternatives, if one is ten times as persistent as another, that will outweigh the alternative being eight times as significant.

To illustrate, suppose that we're in the late nineteenth century and the world is currently on track to use QWERTY keyboards, but if we choose to, we can shift the world to use Dvorak keyboards.[5] In the table below, I'll use X's to represent the course of the counterfactual possible world *p* where we make Dvorak the standard, and I'll use O's to represent the course of the status quo world *q*, where QWERTY is the standard, until time period 4, when Dvorak becomes the standard. After period 4, keyboards are made obsolete by some other technology.

Table A.1. QWERTY vs. Dvorak as Example for the Significance, Persistence, Contingency Framework

	Year 1	Year 2	Year 3	Year 4	Years 5+
DVORAK	X	X	X	⊗	
QWERTY	O	O	O		
OTHER					⊗

We'll assess the state of affairs of "Having Dvorak as the standard." In this example, significance is given by the average increase in value over time from Dvorak being the standard rather than QWERTY, over the time periods (1–3) when the counterfactual state of affairs differs from the status quo.[6] Persistence is given by how long Dvorak would remain the standard if we made it the standard: in this example, it's four time periods. Contingency is given by what proportion of time the counterfactual state of affairs differs from the status quo sequence of states of affairs, over the length of time that the counterfactual state of affairs would persist: in this example it's three-fourths or 75 percent.

This is all defined ex post—without taking uncertainty into account. Given that we never know how significant, persistent, and contingent a state of affairs will be, what we ultimately are interested in is the expected value of SPC, or E(SPC).[7] Note, however, that E(SPC) does not in general equal E(S)E(P)E(C).

We can embed the SPC framework into the ITN framework for prioritising among global problems, which was first proposed by Holden Karnofsky at Open Philanthropy.[8]

In the ITN framework, a global problem is higher priority the more important, tractable, and neglected it is, where these terms can be informally defined as follows:

Importance represents the scale of a problem: How much better would the world be if we solved it?

Tractability represents how easy or difficult it would be to solve the problem.

Neglectedness represents how many resources are already going towards solving the problem.

The SPC framework is closely related to the "importance" dimension—more precisely, the product of significance, persistence, and contingency is proportional to the "importance" term in the version of the ITN framework described below.

One way of formalising the ITN framework is as follows.[9] In this formalisation, it would perhaps be more apt to call it the "importance, tractability, *leverage* framework" because the last factor indicates not how much work is already being done on a problem but rather the effect this prior work has on the cost-effectiveness of further efforts: if there are increasing returns to work, then a problem being less neglected can make further work more cost-effective.

As before, we consider a change from the status quo q to some different world p and the difference this makes regarding a certain state of affairs s. Let S be the amount of progress on a problem represented by the world being in state s—this could, for instance, be the fraction of the total problem that is being solved, or it could be measured according to some intermediate metric such as the number of malaria nets distributed, the number of malaria cases averted, or the numbers of asteroids charted. Let W refer to the amount of work required to bring about the change from q to p (for instance, measured in person-hours, or financial costs in dollars). Finally, let S_0 and W_0 be the total progress and work, respectively, corresponding to the problem being fully solved. We can then define:

Importance $=_{\text{df}} [V_s(p) - V_s(q)] / S$

Tractability $=_{\text{df}} S_0 / W_0$

Neglectedness/Leverage $=_{\text{df}} (S/W) / (S_0/W_0)$

Importance represents how valuable it is to make additional progress on a problem. Tractability represents the average returns if we completely solve the problem. Neglectedness, or leverage, represents how the returns of the specific change under consideration compare to those average returns.

The SPC framework and its relationship to the ITN framework are explained in more depth in a technical report (*The Significance, Persistence, Contingency Framework*, by MacAskill, Thomas, and Vallinder), available on the *What We Owe the Future* book website.

4. Objections to Longtermism

Some objections to longtermism have been discussed in the main text of the book. In particular, I take Chapters 2–7 to address the most obvious objection: that we can't predictably affect the expected value of the long-run future. This appendix discusses other objections to longtermism. More discussion can be found at longtermism.com.

Future People Will Be Better Off

In Chapter 1, I argued against the idea that we should give much less weight to the interests of future people *merely because* they'll live in the future (while allowing for potentially giving them moderately less weight because considerations like partiality and reciprocity apply more strongly to the current and the next few generations).

Economists sometimes give an alternative reason for discounting future impacts: that future people will be richer than we are. A given *economic* benefit will therefore be worth less for future people than it will be for present people, just as £1000 is worth less for a present-day millionaire than it is for someone living in extreme poverty.

This consideration is important as far as it goes. But it can't function as a justification for always giving little weight to the interests of future generations. Future people might well be better off over the next century or two. But whether they will be better off in a thousand years is very unclear; this is especially true when we're worrying about catastrophes like authoritarian takeover, civilisational collapse, or long-run technological stagnation.

Future people's wealth may even be beside the point because the sorts of benefits and harms I'm generally considering look very unlike making some future people slightly richer or poorer. In the case of value lock-in, future people might be just as rich whichever values are locked in; the issue is whether that future wealth is used to create flourishing or misery. In the case of extinction, the issue is whether future people exist at all. In either case, the simplifying assumption that some future harm or benefit just makes future people a little poorer or richer is not accurate, and the fact that future people would be richer than us (if they were to exist) is neither here nor there.

Future People Can Take Care of Their Own Problems

Perhaps we should endorse a division of labour between different generations. There are some problems that we face in our time, which we should take care of. There are some problems that future people will face, which they should take care of.

Even if you're sympathetic to this line of argument in general, I don't think it has any plausibility when it comes to the issues I discuss in this book. In the case of value lock-in, the issue precisely concerns what future people will see as a problem or not: if there is a future dystopia where enslaving people is regarded as entirely acceptable, those in charge of society won't see it as a problem, and we shouldn't expect them to try to change it. In the case of a permanent catastrophe, those in the future cannot undo the effects of our actions; in the case of extinction, there aren't even any future people around!

Consider also that some problems for future people are caused by us, and it's often easier to prevent a problem from occurring than it is to fix the problem once it's happened. It's easier to avoid breaking a glass than it is to piece it back together once it has smashed; it's easier to avoid burning coal than it is to suck carbon dioxide out of the atmosphere.

We Should Not Chase Tiny Probabilities of Enormous Amounts of Value

In this book, I've relied on the idea that, under uncertainty, the value of an action is given by its expected value. But this idea faces problems when we're considering actions that have a tiny probability of success but would have enormous value if successful. For example, suppose that you can either save ten lives for certain or take an action that has a one in a trillion trillion trillion chance of saving one hundred trillion trillion trillion lives. Even though the expected lives saved by the latter action is greater, it seems very intuitive that the right thing to do is to take the safe bet and save the ten lives for certain. Taking the low-probability action seems wrong.

Unfortunately, there is no good solution to this problem; it has been shown that any theory of how to make decisions under uncertainty faces highly unintuitive consequences.[10] If we wish to avoid the idea that tiny probabilities of enormous amounts of value can be better than guarantees of merely large amounts of value, then we run into other problems that seem just as bad.

For the purpose of this book, my response to this problem is simply that, at least in the world as it is today, the probabilities under discussion are not at all tiny. The probability that there will be a civilisation-ending catastrophe over the coming centuries is greater than 0.1 percent; and the probability that civilisation lasts longer than a million years is greater than 10 percent; and there are actions, such as investing into clean-energy R & D or stockpiling protective equipment against future pandemics, that predictably reduce the chance of some catastrophes by a non-tiny amount.

It may well be that the probability of any one individual having an impact on some major event like an existential catastrophe is small. But the same is true for many ordinary sorts of morally motivated actions. If you join a protest, or vote, or sign a petition, then the chance that your action will make a difference to the outcome is very small. Nonetheless, these are actions that we often should take, because the probabilities aren't tiny and the gains are very great if we *do* make the difference.

We Must Respect Constraints Such as Not Violating Rights

A separate objection comes from the idea of constraints on moral action. Couldn't longtermism justify violating rights in pursuit of longterm benefit, or even justify mass atrocities?

Such courses of action do not follow from longtermism. Concern for the environment does not justify bombing power plants, even if doing so would benefit the environment; concern for the rights of women does not justify assassinating political leaders, even if doing so would benefit women. Similarly, concern for the longterm future does not justify violating others' rights, for two reasons.

First, in practice, violating rights is almost never the best way of bringing about positive longterm outcomes. Yes, we can dream up extreme philosophical thought experiments ("Would it be justified to kill baby Hitler?") in which rights violations bring about the best outcomes. But these essentially never arise in real life. There is an enormous amount that we can do to make the long term go better by peaceful means such as persuading others and promoting or implementing good ideas. Doing these things is clearly a better path than anything that might violate others' rights.

Second, if we either endorse nonconsequentialism or take moral uncertainty seriously, we should accept that the ends do not always justify the means; we should try to make the world better, but we should respect moral side-constraints, such as against harming others.[11] So even on those rare occasions when some rights violation *would* bring about better longterm consequences, doing so would not be morally acceptable.

Longtermism Is Too Demanding

A final line of objection to longtermism is that it's too demanding. If we truly were to give the interests of future generations the same weight as our own, then shouldn't we be willing to almost entirely sacrifice the interests of the present in order to provide even further benefits to future people? And isn't that idea absurd?

This objection does point to a difficult philosophical issue: How much *should* we in the present be willing to sacrifice for future generations? I don't know the answer to this. All I've claimed in this book is that concern for the longterm future is at least one key priority of our time. I'm not claiming that everything we do should be in the service of posterity. But it does seem to me that we should be doing much more to benefit future generations than we currently are.

In particular, at the current margin, the "sacrifices" involved in radically increasing concern for the future generally seem very small or even nonexistent. Given how neglected longterm issues currently are, there are many ways of benefiting the longterm future that also have major benefits to the present. Reducing our consumption of fossil fuels has very longterm benefits, but it also reduces air pollution, which alone kills millions of people each year.[12] An extinction-level pandemic would foreclose all possible future human value, but it would also kill everyone alive today; the probability of this, and of other globally catastrophic pandemics, is more than enough to justify taking dramatically more action to prepare against future pandemics than we do today.[13] Like many other longterm-oriented actions, these are win-wins.

Figure Credits and Data Sources

Figure 1.4. Based on a graph by Our World in Data (Ritchie 2020a). Data from Sovacool et al. (2016); Markandya and Wilkinson (2007).

Figure 1.5. Based on a graph by Our World in Data (2017b). Data for years 1–1989 from New Maddison Project Database (Bolt and van Zanden 2020); data for years 1990 and later from World Bank (2021f).

Figure 2.1. Megatherium based on Haines and Chambers (2006); *Notiomastodon* based on Larramendi (2016, 557), dire wolf based on Wikipedia (2021b).

Figure 3.1. Detail of the public-domain image *Stowage of the British slave ship "Brookes" under the Regulated Slave Trade Act of 1788*, provided by the Library of Congress at https://loc.gov/pictures/resource/cph.3a34658/.

Figure 3.3. Based on a graph by Our World in Data (Ortiz-Ospina and Tzvetkova 2017). Data on female labour force participation from International Labour Organization, as published by the World Bank (2021m); data on GDP per capita from World Bank (2021m).

Figure 4.1. Based on a graph by Carbon Brief (Evans 2019), adapted with permission. Data from IEA (2019) and previous editions of the IEA's World Energy Outlook.

Figures 6.1–6.2. Based on population estimates by Morley (2002; for 200 BC and 130 BC) and Morris (2013; for AD 1 and all later dates).

Figure 6.3. Based on a graph by Our World in Data (2019h). Data for 1976–2009 from Lafond et al. (2017); data for later years from IRENA (2020, 2021).

Figure 7.1. Adapted with permission from Figure 1 in Crafts & Mills (2017). Data from Fernald (2014).

Figure 7.3. Based on a graph by Our World in Data (n.d.-c). Data from UN (2019b).

Figure 7.4. Based on a graph by Our World in Data (2020c). Original data sources: Data on live births per woman from UN (2019b); data on GDP per capita from Penn World Table (Feenstra et al. 2015).

Figure 9.1. Based on a graph by Our World in Data (2021d). Data on life satisfaction from Gallup World Poll ("Cantril Ladder" question) as published in the World Happiness Report (Helliwell et al. 2021); data on GDP per capita from World Bank (2021m).

Figure 9.2. Based on a graph by Our World in Data (Ortiz-Ospina and Roser 2017, Section "Economic Growth and Happiness"). Data on happiness from World Values Survey 7 (2020); data on GDP from Penn World Table (Feenstra et al. 2015).

Figure 9.3. Based on a graph by Our World in Data (Roser 2013d, Section "Global Divergence Followed by Convergence"). Data calculation by Ola Rosling for Gapminder (2021) based on multiple sources.

Figure 9.4. Based on a graph by Our World in Data (Roser, Ortiz-Ospina, and Ritchie 2019, Section "Rising Life Expectancy Around the World"). Data on world average pre-1950 from Riley (2005ab); country-level data pre-1950 from Clio Infra, as published by Zijdeman and Ribeira da Silva (2015); data for 1950 and later years from UN (2019b).

Figure 9.5. Based on a graph by Our World in Data (2020g). Data from FAO (2021ab).

Notes

Additional notes are available at whatweowethefuture.com/notes.

To locate references such as "Cotra 2020," consult the online bibliography at whatweowethefuture.com/bibliography.

Introduction

1. This thought experiment comes from Georgia Ray's "The Funnel of Human Experience" (G. Ray 2018). A number of commentators have also pointed me to the popular short story "The Egg" by Andy Weir (2009), which has a similar premise.

2. The idea of the "first human being" is a bit of poetic license: there is no strict dividing line between *Homo sapiens* and our forebears. Moreover, it's not even clear that "we" should refer only to *Homo sapiens*: early humans mated with Neanderthals and Denisovans (L. Chen et al. 2020). These issues do not alter the upshot of this thought experiment.

While the timing of *Homo sapiens*'s speciation is sometimes cited as two hundred thousand years ago, expert consensus is now that it occurred three hundred thousand years ago (Galway-Witham and Stringer 2018; Hublin et al. 2017; Schlebusch et al. 2017; personal communication with Marlize Lombard, Chris Stringer, and Mattias Jakobsson, April 26, 2021).

3. The best available estimate is 117 billion (Kaneda and Haub 2021).

4. These and similar claims are based on combining estimates of the total human population (Kaneda and Haub 2021) and life expectancy at different times (Finch 2010; Galor and Moav 2005; H. Kaplan et al. 2000; Riley 2005; UN 2019c; WHO 2019, 2020). They should be treated as ballpark estimates.

5. These numbers, which I've based on back-of-the-envelope calculations, are meant to be merely illustrative. The true figures, if we had them, would probably be slightly different from what I've used here. More at whatweowethefuture.com/notes.

6. Slavery is absent today among what are (erroneously) known in the literature as socially "simple," highly egalitarian hunter-gatherer societies, who are probably most similar to preagricultural human societies (Kelly 2013, Chapter 9). Slavery likely only became widespread after the emergence of sedentary societies following the agricultural revolution. Any estimate of the fraction of the population enslaved since then necessarily involves some guesswork. But the evidence that exists suggests that in many agricultural societies, around 10 to 20 percent of the population was enslaved. For example, in the second millennium AD, as much as one-third of the population of Korea was enslaved. A quarter to a third of the population of some areas of Thailand and Burma were enslaved in the seventeenth through the nineteenth centuries and in the late nineteenth and early twentieth centuries, respectively. The enslaved population of the city of Rome during the Roman Empire was estimated to be between 25 and 40 percent of the total population. Probably around a third of people in ancient Athens were enslaved. In 1790, approximately 18 percent of the American

population was enslaved (Bradley 2011; Campbell 2004, 163; Campbell 2010; D. B. Davis 2006, 44; Hallet 2007; Hunt 2010; Joly 2007; Patterson 1982, Appendix C; J. P. Rodriguez 1999, 16–17; Steckel 2012). Slavery was abolished globally over the course of the nineteenth and twentieth centuries.

Estimating the fraction of the population who owned enslaved people involves equal amounts of guesswork, but it is reasonable to think that the proportion of slaveholders was similar to the proportion of the enslaved. If one-quarter of the population in a society were enslaved, then one might reasonably guess that they were owned by the richest quarter of the society. For instance, in America in 1830, there were around two million enslaved people and, according to one survey, 224,000 slaveholders in the South. However, this assumes that only one person in a surveyed household should be considered a slave owner, but arguably we should count everyone in the whole household. Since the household likely would have included more than five people, this suggests that there were around two enslaved people per slave owner (R. Fry 2019; Lightner and Ragan 2005; O'Neill 2021b). And the US South probably had an historically high ratio of slave owners to enslaved people.

7. See whatweowethefuture.com/notes.

8. In this thought experiment as I currently state it, you would live to the end of the lives of all those alive today, but not beyond. I am taking into account today's greater life expectancy—if we only looked at the number of people, ignoring how long they live, then current people account for 7 percent of those who have ever lived (Kaneda and Haub 2021). If, for people currently alive, we only included their experience until the present moment—rather than until the expected end of their lives—their share of all experience would be closer to 6 percent, since many people have long lives ahead of them.

9. See whatweowethefuture.com/notes.

10. "Seconds" is about accurate if we maintain roughly the current population as long as Earth remains habitable. If we settle other solar systems or otherwise massively increase either the population or the life span of civilisation, then really it should be tiny fractions of seconds. It is not out of the question that the experience of all past and present people could correspond to a time interval that is shorter than the shortest one ever measured—2.47 zeptoseconds, or 2.47×10^{-19} seconds (Grundmann et al. 2020), many orders of magnitude less time than it would take for your eyes to chemically react to light before initiating a neural transmission (Weiner 2009). This would be the case if, for instance, for a hundred trillion years (until the end of the age of star formation) each of one hundred billion stars (the lower bound of typical estimates for the number of stars in our galaxy, the Milky Way) supported a population of ten billion people (approximately the current world population).

11. Throughout this book, I drop the hyphen and use "longterm" as an adjective. I use "long term" as the noun phrase.

12.See https://www.givingwhatwecan.org/.

Chapter 1: The Case for Longtermism

1. This example is modified from *Reasons and Persons* (Parfit 1984, 315).

2. Though this is sometimes described as an ancient Chinese or ancient Greek proverb, its origin is unknown.

3. *Constitution of the Iroquois Nations* 1910.

4. Lyons 1980, 173.

5. That said, some reciprocity-type reasons might motivate concern for future generations, too. We may not benefit from the actions of people in the future, but we benefit enormously from the actions of people in the past: we eat fruit from plants they bred over thousands of years; we rely on medical knowledge they developed over centuries; we live under legal systems shaped by countless reforms they fought for. Perhaps, then, this gives us reasons to "pay it forward" and do our part to benefit the generations to come.

6. In the famous "to be, or not to be" soliloquy from *Hamlet*, "undiscovered country" refers to the afterlife: "But that the dread of something after death, / The undiscovered country from whose bourn / No traveller returns, puzzles the will / And makes us rather bear those ills we have / Than fly to others that we know not of?" In appropriating (and naturalizing) that metaphor to refer instead to the future, I'm following the lead of the Klingon chancellor Gorkon from the eponymous *Star Trek VI: The Undiscovered Country*.

7. Common estimates are 2.5 million (Strait 2013, 42) to 2.8 million years (DiMaggio et al. 2015).

8. Özkan et al. 2002, 1797; Vigne 2011. More on the formation of the first cities online.

9. Barnosky et al. 2011, 3; Lawton and May 1995, 5; Ord 2020, 83–85; Proença and Pereira 2013, 168.

10. I don't mean to make any strong claim that no nonhuman animals possess any abstract reasoning or longterm planning abilities whatsoever, or that none of them use any tools. There is ample evidence for several species arguably planning hours or even days ahead (e.g., Clayton et al. 2003; W. A. Roberts 2012), and tool production and use in apes is well documented (Brauer and Call 2015; Mulcahy and Call 2006). More broadly, animal cognition is a topic of ongoing empirical research and lively philosophical debate (for an overview, see Andrews and Monsó 2021).

11. Estimates of how long the sun will continue to burn range from 4.5 billion (Bertulani 2013) to 6.4 billion years (Sackmann et al. 1993), though 5 billion seems to be the most common rough figure. More precisely, this refers to the time by which all hydrogen in the sun's core will be used up, at which point the sun will begin to leave what astronomers call the "main sequence" of stars. However, it is still going to "burn"—that is, to generate energy through nuclear fusion of hydrogen into helium, albeit in its shell rather than its core. After it expands as a red giant for about two to three billion years, nuclear fusion is going to resume in the core—this time fusing helium into carbon and oxygen—and only after this final helium flash will the sun stop shining altogether, about eight billion years into the future.

The figure for conventional star formations is from F. C. Adams and Laughlin 1997, 342.

I am grateful to Toby Ord for making me aware of how long a few stars will continue to shine. Anders Sandberg, in his upcoming book *Grand Futures*, notes that on even longer timescales, after the end of those stars, there are more exotic sources of energy, such as black holes, which could be harnessed. This could extend civilisation's life span beyond a million trillion years.

12. Wolf and Toon (2015, 5792) estimate that "physiological constraints on the human body imply that Earth will become uninhabitable for humans in ~1.3 Gyr [1.3 billion years]"; Bloh (2008, 597) gives a somewhat shorter window, stating that the "life spans of complex multicellular life and of eukaryotes end at about 0.8 Gyr and 1.3 Gyr from present, respectively." I am going with a more conservative window of human habitability of perhaps five hundred million years because of considerable uncertainty about the timing and likelihood of key developments—such as plants dying from carbon dioxide starvation, or a "runaway greenhouse effect" leading to the evaporation of the oceans—and the open question of which of these will be the limiting factor for human habitability (see Heath and Doyle [2009] for a survey of considerations that affect the habitability of planets for different types of life). More at whatweowethefuture.com/notes.

13. See whatweowethefuture.com/notes.

14. There are one hundred to four hundred billion stars in our galaxy, the Milky Way. The number of reachable galaxies has been estimated as 4.3 billion by Armstrong and Sandberg (2013, 9) while Ord (2021, 27) states, "The affectable universe contains about 20 billion galaxies with a total of between 10^{21} and 10^{23} stars (whose average mass is half that of the Sun)."

15. My figures are for life expectancy at birth (Roser 2018). Since, in the early nineteenth century, about 43 percent of children globally died before age five (Roser 2019), someone surviving until that age could expect to become about fifty years old. Note also that seventy-three years is not necessarily the best prediction for how long someone born today is going to live: the figures I quoted are for what's known as "period life expectancy," a measure of life expectancy that by definition ignores future trends. For instance, if there will be further progress in medicine and public health, then someone born today should in fact expect to live longer than seventy-three years; on the other hand, if new deadly diseases will emerge or a large fraction of the world population will be wiped out by a large-scale catastrophe, someone born today should expect to live a shorter life than suggested by their period life expectancy at birth.

16. In 1820, an estimated 83.9 percent of the world population lived on a daily income that, adjusted for inflation and price differences between countries, bought less than one dollar did in the US in 1985 (Bourguignon and Morrisson 2002, Table 1, 731, 733). In 2002, when Bourguignon and Morrisson published their seminal paper on the history of the world income distribution, this was the World Bank's international poverty line, typically used to define extreme poverty. The World Bank has since updated the international poverty line to a daily income corresponding to what $1.90 would have bought in the US in 2011. Using this new definition, World Bank data indicates that the share of the global population living in extreme poverty has been less than 10 percent since 2016; the COVID-19 pandemic tragically broke the long-standing trend of that percentage declining year after year, but it did not quite push it over 10 percent again (World Bank 2020). While the extent to which the old and new poverty lines match is often debated, I think the conclusion that the share of the world population in extreme poverty declined dramatically is unambiguous. This is not to deny we still have a long way to go in the fight against poverty; for instance, more than 40 percent of the world population still live on less than $5.50 per day (again, adjusted for inflation and international price differences relative to the US in 2011).

17. Roser and Ortiz-Ospina 2016.

18. Our World in Data 2017a. More at whatweowethefuture.com/notes.

19. There are a few rumoured cases of women being awarded degrees or teaching at universities prior to 1700, but their lives are usually poorly documented. More at whatweowe thefuture.com/notes.

20. "Throughout the eighteenth century and up until 1861, all penetrative homosexual acts committed by men were punishable by death" (Emsley et al. 2018).

21. "At the end of the eighteenth century, well over three quarters of all people alive were in bondage of one kind or another, not the captivity of striped prison uniforms, but of various systems of slavery or serfdom" (Hochschild 2005, 2). The numbers for today—40.3 million, or about 0.5 percent of the world population—include both forced labour and forced marriage (Walk Free Foundation 2018).

22. While the broad trend of increasing political liberties and individual autonomy strikes me as incontrovertible, the exact numbers depend on the definition of democracy. I got mine from Our World in Data's page on "Democracy" (Roser 2013a), which is based on the widely used Polity IV data set. Its democracy score is a composite variable that captures different aspects of measuring "the presence of institutions and procedures through which citizens can express effective preferences about alternative policies and leaders" and "the existence of institutionalized constraints on the exercise of power by the executive" but excludes measures of civil liberties (Marshall et al. 2013, 14). My claim about the year 1700 is based on the assumption that the situation then can't have been much better than in the early nineteenth century, when Polity IV has less than 1 percent of the world population living in a democracy. I'm also making the definitional judgment call to exclude societies without full-blown statehood (e.g., hunter-gatherers) even if some of them might have had

protodemocratic features such as inclusive participation in deliberation or checks on leaders' ability to abuse power.

23. Gillingham 2014, Wyatt 2009. In total, the British Empire bought more than three million enslaved people during the transatlantic slave trade, and France bought more than one million (Slave Voyages 2018).

24. Sonnets 1–126 are typically considered to be addressed to a "young man," though, like many aspects of Shakespeare's life and works, this remains a subject of scholarly debate. More at whatweowethefuture.com/notes.

25. Shakespeare 2002, 417.

26. Shakespeare "had likely drafted the majority of his sonnets in 1591–95" (Kennedy 2007, 24). Kennedy cites Hieatt et al. (1991, 98) who, based on an analysis of rare words appearing in Shakespeare's works throughout his career, specifically suggest that "many of" Sonnets 1–60 were first drafted between 1591 and 1595.

27. See whatweowethefuture.com/notes.

28. Horace 2004, 216–217.

29. See whatweowethefuture.com/notes.

30. See whatweowethefuture.com/notes.

31. The quote is from Rex Warner's 1954 translation as printed in the 1972 Penguin Books edition (Thucydides 1972). More at whatweowethefuture.com/notes.

32. Bornstein 2015, 661; Holmes and Maurer 2016. More at whatweowethefuture.com/notes.

33. J. Adams 1851, 298. Incidentally, in the same preface, Adams quotes Thucydides at length, including part of the passage I referenced earlier.

34. My rendition of how Franklin's will came about employs some interpretative best guesses. More at whatweowethefuture.com/notes.

35. Franklin's bequest is well known. My source for the numbers given in the main text is the epilogue of Isaacson (2003, 473–474). More at whatweowethefuture.com/notes.

36. See whatweowethefuture.com/notes.

37. Lloyd 1998, Chapter 2.

38. Lord et al. 2016; Talento and Ganopolski 2021. Of course, we might later remove carbon dioxide from the atmosphere. But we should not be very confident that we will do this, and certainly not in light of the possibilities of collapse and stagnation that I discuss in Chapters 6 and 7. I discuss the longtermist importance of burning fossil fuels in more detail in Chapter 6.

39. Hamilton et al. 2012.

40. The average life span of carbon dioxide shows another way in which current climate rhetoric and policy is shortsighted: the comparison with methane. Methane is often claimed to have thirty or even eighty-three times the warming potential of carbon dioxide, or even more. But from a longterm perspective, these numbers are misleading. Methane only stays in the atmosphere for about twelve years (IPCC 2021a, Chapter 7, Table 7.15); this is in stark contrast to carbon dioxide, which, as we've seen, stays in the atmosphere for hundreds of thousands of years.

The most commonly used weighting for methane has been to treat it as thirty times as important as carbon dioxide, but this metric measures the effect methane has on temperatures after forty years. (Confusingly, this metric is known as "Global Warming Potential.") If instead we measure the effect that methane has on temperatures in one hundred years, methane is only 7.5 times as potent as carbon dioxide (IPCC 2021a, Chapter 7, Table 7.15).

Though the weight we give to methane rather than carbon dioxide is usually presented as a scientific matter, really it's primarily about whether we wish to prioritise reducing climate change over the next few decades or over the long run (Allen 2015). Given that we emit sixty times as much carbon dioxide as methane, if we take a longterm perspective, it's carbon dioxide that should be our main focus (H. Ritchie and Roser 2020a; Schiermeier 2020).

41. P. U. Clark et al. 2016.

42. IPCC 2021a, Figure SPM.8. The medium-low-emissions scenario is known as RCP4.5 (Hausfather and Peters 2020; Liu and Raftery 2021; Rogelj et al. 2016).

43. Clark et al. (2016, Figure 4a) project that on a medium-low-emissions scenario, sea level would rise by twenty metres. Van Breedam et al. (2020, Table 1) find that sea level would rise by ten metres on the medium-low pathway.

44. P. U. Clark et al. 2016, Figure 6.

45. See whatweowethefuture.com/notes.

46. Our World in Data 2020a, based on Lelieveld et al. 2019. This only includes deaths from outdoor air pollution. An additional 1.6 million (Stanaway et al. 2018) to 3.8 million (WHO 2021) excess deaths per year are due to indoor air pollution, much of which is caused by lack of access to electricity and clean fuels for cooking, heating, and lighting (H. Ritchie and Roser 2019). More than 2.5 billion people are able to cook only by burning coal, kerosene, charcoal, wood, dung, or crop waste using inefficient and unsafe technology such as open fires (WHO 2021).

47. "In Europe an excess mortality rate of 434 000 (95% CI [confidence interval] 355 000–509 000) per year could be avoided by removing fossil fuel related emissions. . . . The increase in mean life expectancy in Europe would be 1.2 (95% CI [confidence interval] 1.0–1.4) years" (Lelieveld, Klingmüller, Pozzer, Pöschl, et al. 2019, 1595). A 95 percent confidence interval indicates the range in which, based on the authors' model, the true number falls with a probability of 95 percent. Note that the authors use spacing rather than commas when formatting large numbers—e.g., "434 000" refers to four hundred thirty-four thousand.

48. Scovronick et al. (2019, 1) found that depending on air-quality policies and "on how society values better health, economically optimal levels of mitigation may be consistent with a target of 2°C or lower." Markandya et al. (2018, e126) found that the "health co-benefits substantially outweighed the policy cost of achieving the [2°C] target for all of the scenarios that we analysed" and that "the extra effort of trying to pursue the 1.5°C target instead of the 2°C target would generate a substantial net benefit in India (US$3.28–8.4 trillion) and China ($0.27–2.31 trillion), although this positive result was not seen in the other regions."

49. The claim that we live in a highly unusual period in history also raises some interesting philosophical issues, as I discuss in my article "Are We Living at the Hinge of History?" (for a draft see MacAskill 2020, formal publication forthcoming). However, note that the arguments in that article are against the idea that we're at the *most* influential time ever. I think the case for thinking that we're ("merely") at an enormously influential time is very strong.

50. This argument and framing follows Holden Karnofsky's "This Can't Go On" (2021b), which builds on an argument by Robin Hanson (2009). Further discussion at whatweowethefuture.com/notes.

51. More precisely, I'm thinking of the present as a postindustrial era that began 250 years ago and will end whenever growth rates slow again to below 1 percent per year. For recent growth rates, see World Bank (2021e).

52. For all claims about the history of global growth, see, for instance, DeLong (1998). For an overview of other data sources, which give similar numbers, see Roodman's (2020a) data and Roser's (2019) data sources. Note that my claims are about average growth rates that are being sustained for several doubling times—we cannot, of course, rule out that the growth rate may have been 2 percent in a single year in, say, 200,000 BC (but we know that, if this happened, it must have been an exception). For a discussion of intermittent brief

periods of above-average growth in world history, see Goldstone (2002), though my background research for Chapter 7 suggests that some examples therein are controversial.

53. Energy use: Our World in Data 2020f; carbon dioxide emissions: Ritchie and Roser 2020a; land use: Our World in Data 2019b. Measurements of scientific advancement are subject to interpretation, but I believe that few would disagree with the claim that the pace of technological innovation has rapidly accelerated since the Scientific Revolution in the sixteenth century compared to premodern times.

54. This is in fact closer to what growth has been at the technological frontier—that is, ignoring the transient catch-up growth of poorer countries (Roser 2013b).

55. Karnofsky 2021b, nn7–8.

56. For further discussion about whether it's possible, see Hanson 2009 and Karnofsky 2021c.

57. I thank Carl Shulman for this point.

58. See whatweowethefuture.com/notes.

59. Scheidel (2021, 101–107) provides a summary of historic empires' population sizes; his Table 2.2 (103) indicates that the Western Han dynasty comprised 32 percent of the world's population in AD 1, while in AD 150 30 percent lived in the Roman Empire. There is, however, considerable uncertainty about historic population sizes; more at whatweowethefuture.com/notes. The historian Peter Bang (2009, 120) has commented that even at their peak, the Han and the Roman Empires "remained hidden to each other in a twilight realm of fable and myth."

60. This treats the orbit of the outermost planet, Neptune, as the boundary of the solar system. More at whatweowethefuture.com/notes.

61. See whatweowethefuture.com/notes.

62. See whatweowethefuture.com/notes.

63. "Eventually space will expand so quickly that light cannot travel the ever-expanding gulf between our Local Group and its nearest neighbouring group (simulations suggest that this will take around 150 billion years)" (Ord 2021, 7).

Chapter 2: You Can Shape the Course of History

1. Megafauna are technically defined as animals weighing more than forty-four kilograms (Haynes 2018).

2. Technically, glyptodonts are a clade (Zurita et al. 2018).

3. Some larger glyptodonts weighed 1.5 tonnes (Delsuc et al. 2016), which is more than a Ford Fiesta. Towards the end of the Pleistocene, many glyptodonts weighed more than two tonnes and were five metres long (Defler 2019b).

4. This was true of *Doedicurus*, one genus of glyptodont (Delsuc et al. 2016).

5. It is always difficult to estimate exactly when a species went extinct, for several reasons. In the case of the glyptodonts, there is significant debate about the dating of certain fossils, with some estimates suggesting their last appearance dates to only seven thousand years ago, though there are concerns about the reliability of these estimates (Politis et al. 2019). The latest uncontroversial radiocarbon-dated glyptodont bone suggests a last-appearance date of 12,300 years ago. However, glyptodont bones have been recovered in strata that have been dated to 12,000 years ago, and maybe later (Barnosky and Lindsey 2010; Prado et al. 2015, Table 2; Ubilla et al. 2018).

6. Defler 2019a, xiv–xv. Some scholars think that megatherium was bipedal, though this is controversial. If so, it was the largest bipedal mammal ever (Amson and Nyakatura 2018).

7. Some earlier estimates suggested that megatherium might have lived into the Holocene, but recent work has put the last-appearance date of megatherium at around 12,500 years ago (Politis et al. 2019). Because of the patchiness of the fossil record, the latest fossil

of a species that we've found is probably not the very last individual of a species. This is known as the Signor-Lipps effect.

8. Mothé et al. 2017, Section 3.5; 2019. Electron spin resonance dating of bones is less reliable than radiocarbon dating of collagen, and the last-appearance date of *Notiomastodon* is highly controversial (Dantas et al. 2013; Oliveira et al. 2010, Table 2). Thanks to Emily Lindsey (personal communication, November 22, 2021) for discussion of this point.

9. The dire wolf weighed around 68 kilograms, with a maximum weight of 110 kilograms (Anyonge and Roman 2006, Table 1; Sorkin 2008). The dire wolf is a member of the Caninae subfamily and is therefore a canine, but recent research has shown that it is not actually a wolf: although it looks similar to the grey wolf, this is a case of convergent evolution (Perri et al. 2021). The largest member of the Canidae family, of which Caninae is a subfamily, was *Epicyon haydeni*, which weighed up to 170 kilograms. As with all megafauna, the precise reason that the dire wolf became extinct is disputed. More online.

10. For a review of the case for the anthropogenic explanation, see, for example, Haynes (2018), Koch and Barnosky (2006), Surovell and Waguespack (2008), Smith et al. (2019), and Wignall (2019b). The two main pieces of evidence in favour of a central role for humans are as follows. First, the megafaunal extinctions in particular regions all happened after or around the time of the first recorded human arrival in those regions. Some of the last fossils for the extinct genera appear before the first human fossil, but this is probably due to gaps in the fossil record. Second, the extinctions were highly skewed towards easy-to-hunt big animals, which would have been especially valuable to human hunters. The extent of the skew is wholly unique for species extinctions in the last sixty-six million years.

For arguments supporting mostly natural causes, see Meltzer (2015, 2020) and Stewart et al. (2021). There are two main arguments against a leading role for humans. First, some argue that the number of kill sites is too low given the scale of megafaunal slaughter that would have been required. However, proponents of the anthropogenic theory argue that given the patchiness of the fossil record, the number of identified megafaunal kill sites is actually large in a paleontological context, and that absence of evidence is not evidence of absence. Second, some argue that the earliest people are unlikely to have been sufficiently abundant or technologically sophisticated to kill millions of megafauna. However, modelling evidence suggests that humans probably were numerous enough to cause extinctions on the scale suggested.

The main problems for the climate change explanation are as follows. First, in addition to the transition out of the Pleistocene, megafauna lived through many dramatic climate changes over the last few million years. In North America, for example, the vast majority of the extinct genera lived through more than twelve glacial-interglacial cycles that were similar to the one at the end of the Pleistocene. Yet it was only at the end of the Pleistocene, when humans were present, that the rates of megafaunal extinction increased so greatly. Second, the climate change theory also struggles to explain the skew towards large mammals. As Wignall (2019b, 107) notes, "Under the normal 'rules' of extinction, highest losses generally occur among species with a relatively limited habitat range, but the Pleistocene extinctions were fundamentally different. Many of the megafaunal species inhabited a vast geographic extent: the woolly mammoth and woolly rhino ranged across the whole of Eurasia and North America." Finally, the climatic changes that megafauna were exposed to across different continents were very different—in some cases cooling, in others warming, in others drying, and so on—and yet they uniformly led to megafaunal extinctions across different ecological niches.

For arguments that both humans and natural causes contributed to the extinction of megafauna, see Broughton and Weitzel (2018) and Metcalf et al. (2016).

11. In only the last eight hundred thousand years, there have been eleven glacial-interglacial transitions, many of which seem similar to the Pleistocene-Holocene transition

(PAGES 2016). Earlier in the Pleistocene, glacial-interglacial transitions were more frequent but less dramatic (Hansen et al. 2013). Most of the megafauna evolved millions of years ago, so they had to survive more than a dozen such transitions (Meltzer 2020).

12. Koch and Barnosky 2006; S. K. Lyons et al. 2016.

13. F. A. Smith et al. 2019. Human fossils do not always overlap with the fossils of extinct species. This is plausibly explained by the patchiness of the fossil record and the Signor-Lipps effect. For discussion, see Meltzer (2020) and Haynes (2018).

14. Varki 2016; Wignall 2019b.

15. J. O. Kaplan et al. 2009, Table 3; Stephens et al. 2019; Zanon et al. 2018, Figure 10.

16. The IPCC *Fifth Assessment Report* estimates that preindustrial land-use change increased carbon dioxide concentrations by around ten parts per million, which would have caused a warming of 0.16 degrees (assuming a climate sensitivity of three degrees; IPCC 2014a, Section 6.2.2.2). The IPCC's 2021 *Sixth Assessment Report* does not quantify the effects of preindustrial land-use change, but it seems to suggest that the role of land-use change in increasing carbon dioxide concentrations is small relative to natural changes (IPCC 2021a, Section 5.1.2.3). Others argue that the human preindustrial contribution was much larger and may even have prevented an ice age (Ruddiman et al. 2020).

17. This framework was created by Aron Vallinder and me and further developed by Teruji Thomas. It's described more precisely in Appendix 3. It fits nicely with the "importance, tractability, and neglectedness" framework which is widely used in effective altruism when prioritising among causes. The SPC framework provides a way of estimating a quantity proportional to the "importance" dimension.

18. In this framework, it's helpful to assume an end date of the universe; otherwise we would have to deal with some states of affairs being infinitely persistent. We could specify the end of the universe as, for example, the time at which the last black hole disappears from the currently affectable universe.

19. Revive and Restore, n.d.

20. The term "trajectory change" was first coined by Nick Beckstead (2013). In his initial definition, a trajectory change was any very long-lasting or permanent change to the value of the world. With his permission, I've narrowed this definition so that "trajectory change" refers just to long-lasting changes to the average value of civilisation over time, rather than encompassing changes to civilisation's duration too.

21. I am not claiming that I give an exhaustive account of all the ways to positively influence longterm value. A full discussion would at least include the preservation of information (such as historical records, records of languages and cultures, and records of species' genetic makeup) and changes to political institutions, both of which seem important from a longterm perspective.

22. Throughout this book, I focus on scenarios that I think are of particularly great importance from a longterm perspective, like value lock-in and extinction. I don't often say precisely how likely I think these scenarios are, or precisely how valuable I think it is to avoid them. This note gives an overview of my views. I present these views primarily so that engaged readers can understand my views in the context of others', and to explain why I've focused on what I focus on. But I'll offer these caveats: First, they come with extraordinary amounts of uncertainty; I think that one could very reasonably have very different views than I do. Second, though I've tried to be as precise as I can, many of the claims I give credence to are still vague. Third, my credences (that is, my subjective probability estimates) are very likely to change as I get more evidence and my views evolve. Even by the time this book is published, I will probably disagree with several of the numbers I give here.

This century (between now and 2100), the world could take one of approximately four trajectories. Global GDP could continue to grow at approximately the same rate (2–4 percent annually) as it has for the last hundred years. Or it could grow even faster, perhaps driven by advances in artificial intelligence. Or it could grow somewhat slower, tending towards stagnation. Or there could be a major global catastrophe that results in billions dead. I think that the likelihood of each of these four scenarios is between 10 percent and 50 percent. I think that the stagnation scenario is most likely, followed by the faster-than-exponential growth scenario, followed by continued-exponential scenario, followed by the catastrophe scenario. If I had to give precise credences, I'd say: 35 percent, 30 percent, 25 percent, 10 percent.

I think that the chance of value lock-in occurring at some point in time, assuming that civilisation doesn't end before then and not assuming that the lock-in is of a single value system, is greater than 80 percent. I think there's a greater than 10 percent chance of value lock-in happening this century.

I think the total risk of the end of civilisation this century is between 0.1 percent and 1 percent, with most of that risk coming from engineered pathogens, automated weaponry (which I didn't have space to discuss in this book), and currently unknown technology. This doesn't include the possibility of artificial intelligence systems that are misaligned with human preferences taking control of civilisation; I put that possibility at around 3 percent this century, though I'll note that what counts as "misaligned with human preferences" feels vague to me. I think most of the risk we face comes from scenarios where there is a hot or cold war between great powers.

My credence that there will be a catastrophe this century that moves us back to pre-industrial levels of technology is around 1 percent. My credence on recovery from such a catastrophe, with current natural resources, is 95 percent or more; if we've used up the easily accessible fossil fuels, that credence drops to below 90 percent.

I think that the expected value of the continued survival of civilisation is positive, but it's very far from the best possible future. If I had to put numbers on it, I'd say that the expected value of civilisation's continuation is less than 1 percent that of the best possible future (where "best possible" means "best we could feasibly achieve"). Given this credence, trajectory changes have over one hundred times greater potential upside than civilisational safeguarding, though it's often less clear how to confidently make progress when it comes to trajectory changes.

I think there's a lot that we still don't know or understand, including crucial considerations which could dramatically change what we think are top priorities. This makes me feel more positive about building up resources in order to take action in decades' time, rather than trying to take action immediately (e.g., by working on policy around artificial intelligence that is relevant only if artificial general intelligence comes soon). In particular, it makes me feel comparatively positive about building a movement of careful, humble, altruistically motivated people who are trying to figure out how best to improve the world over the long term.

It also makes me feel more positive about taking actions that seem good across a wide variety of worldviews, even if those actions have lower expected value than some other action, on a naive calculation of expected value. (I think that expected value theory is the correct decision theory, at least if we put to the side the "tiny probabilities of enormous amounts of value" problem; my recommendation to sometimes choose actions of seemingly lower expected value is about how we, with our cognitive limitations, should best try to follow expected value theory in practice.) I've held up clean technology and keeping fossil fuels in the ground as examples of this. Other examples would include building bunkers to help humanity weather global catastrophes, reducing the risk of a great-power war, and, again, building a movement of careful, humble, and altruistically motivated people.

My friend and colleague Toby Ord has prominently given a list of estimates of existential risks, which are risks that threaten the destruction of humanity's longterm potential. He puts total existential risk this century at about one in six, with the risk of engineered pandemics at one in thirty and unforeseen anthropogenic risks at one in fifty; he also emphasises that these estimates involved great uncertainty. Our worldviews are broadly very similar, but there are some differences. I put the risks from artificial intelligence and engineered pathogens a bit lower than he does. I am comparatively much more concerned by the lock-in of bad human values than I am of misaligned artificial intelligence takeover. I am more concerned about a great-power war than he is. I think technological stagnation is more likely than he does. I see these differences as "inside baseball"; we hope to get greater clarity on them in the coming years.

The biggest difference between us regards how good we expect the future to be. Toby thinks that, if we avoid major catastrophe over the next few centuries, then we have something like a fifty-fifty chance of achieving something close to the best possible future. I think the odds are much lower. Primarily for this reason, I prefer not to use the language of "existential risk" (for reasons I spell out in Appendix 1) and prefer to distinguish between improving the future conditional on survival ("trajectory changes," like avoiding bad value lock-in) and extending the life span of civilisation ("civilisational safeguarding," like reducing extinction risks). We both agree that how good we should expect the future to be, conditional on no major catastrophe in the next few centuries, is an extremely underexplored issue.

23. See whatweowethefuture.com/notes.

24. Mauboussin, n.d.; Mauboussin and Mauboussin 2018. When stating the range of how subjects interpreted these phrases, I am referring to the fifth and ninety-fifth percentiles of subjects' responses.

25. In a since-declassified memo presented to President Kennedy and Secretary of Defense Robert McNamara by the Joint Chiefs of Staff, it is written that "timely execution of this plan has a fair chance of ultimate success" (Lemnitzer 1961, no 1q). It has been widely cited that "fair chance" corresponded to a roughly 30 percent chance of success (see, e.g., Tetlock and Gardner 2016). This was first reported by journalist Peter Wyden in the book *Bay of Pigs: The Untold Story* (1979) based on interviews with participants. The estimated probability is attributed to Brigadier General David Gray: "When they discussed what 'fair' meant, Gray said he thought the chances were thirty to seventy" (Wyden 1979, 89).

26. See, for example, Koonin 2014.

27. Researchers who have made this point include John Quiggin in "Uncertainty and Climate Change Policy" (Quiggin 2008), Martin L. Weitzman in "Fat-Tailed Uncertainty in the Economics of Catastrophic Climate Change" (Weitzman 2011), and Robert S. Pindyck in "Climate Change Policy: What Do the Models Tell Us?" (Pindyck 2013).

28. The most likely scenario now appears to be around the IPCC's medium-low-emissions scenario, known as RCP4.5 (Climate Action Tracker 2021; Hausfather 2021a; Hausfather and Peters 2020; Liu and Raftery 2021, Figure 1).

29. This probability range is from IPCC (2021a, Table SPM.1).

30. We should be careful to bear in mind that expected SPC does not equal expected S × expected P × expected C. For our purposes, this consideration will not be hugely important.

31. M. Fry 2013.

32. Seth 2011, 305–308.

33. See whatweowethefuture.com/notes.

34. For the history of the writing of the US Constitution, see US National Archives (2021). For a list of constitutional amendments and the date they were passed, see *Encyclopedia Britannica* (2014).

35. The three Civil War amendments had other important effects as well, including serving as the basis for the legal doctrine of incorporation, according to which many parts of the Bill of Rights are binding for state and local governments (rather than just the federal government).

36. See whatweowethefuture.com/notes.

37. See, for example, Zaidi and Dafoe 2021.

38. These texts are discussed in Chapter 11 of John Barton's *A History of the Bible: The Book and Its Faiths* (2020) and include additional gospels, various Gnostic texts, and a set of texts called the Apostolic Fathers. Several versions of the early Christian Bible include additional texts.

39. When precisely the New Testament as we know it was solidified is difficult to establish given the lack of surviving records from the time. However, the Codex Sinaiticus, a fourth-century Greek Bible, includes books called Barnabas and The Shepherd, which are absent from today's New Testament (Barton 2020, Chapter 11).

40. Sherwood 2011; Lapenis 1998. Arrhenius's contribution was notable for its quantitative predictions. The idea that atmospheric greenhouse gas concentrations could affect the climate had been proposed even earlier, in 1864, by physicist John Tyndall. It's also worth noting, however, that Arrhenius reportedly thought the warming would be a good thing, on balance, because Europe would have a milder climate (Sherwood 2011, 38).

41. Capra 2007.

42. *New York Times* 1956. More details on the article are in Kaempffert (1956).

43. NPR 2019.

44. NPR 2019. "Seem to impinge" in original shortened to "impinge" for conciseness.

Chapter 3: Moral Change

1. It's difficult to define "slavery." In my view, there is a spectrum of economic arrangements under which a worker can be more or less free, in many different ways, and there is no precise set of such arrangements that deserve to be called "slavery." In this chapter, by "slavery" I mean an economic arrangement where people are so unfree as to be in some significant ways treated as property, even if this is not recognised in the law. I mean this to include not just transatlantic chattel slavery but also slavery as historically practised in Europe, India, China, Africa, the Arabic world, the Americas, and so on. I exclude serfdom and indentured servitude from my definition.

2. The prevalence of slavery in early agricultural civilisations is well established among reference works (Egypt: Allam 2001; India: Levi 2002; Mesopotamia: Reid 2017; China: Yates 2001).

3. Eltis and Engerman 2011, 4–5. Some data on why people were enslaved comes from a survey conducted by Sigismund Wilhelm Koelle, a linguist who surveyed people in Sierra Leone while employed by the Church Missionary Society between 1847 and 1853. This is discussed in Curtin and Vansina (1964).

4. Estimates of slavery's historical prevalence are highly uncertain, even for relatively well-documented societies like Rome's. But most estimates suggest that 10 percent is a reasonable lower bound. Walter Scheidel (2012, 92) estimates a range of 5 percent to 20 percent, with a best guess of 10 percent, while Harper (2011, 59–64) estimates it was "on the order of" 10 percent for the later empire (AD 275–425). Patterson (1982, 354) gives a higher estimate of 16–20 percent between the years of AD 1 and 150.

5. Campbell 2010, 57; Ware 2011.

6. Rudolph T. Ware III writes that the "best scholarly estimate" of the number of enslaved people taken from sub-Saharan Africa in the "so-called Arab trade" between AD 650 and 1900 is "roughly 11.75 million" (Ware 2011, 51). But this estimate is highly uncertain and does not account for people enslaved in Central Asia or Europe, nor for people enslaved and traded within sub-Saharan Africa. The true figure for the total number of enslaved people exported across the Sahara or Indian Ocean could be somewhat lower, or much higher, than twelve million.

7. These numbers come from the Slave Voyages database (Slave Voyages 2018).

8. "Most historians rightly assert that warfare was at the core of slaving and that most of the enslaved Africans shipped to the Americas were captives of war" (Ferreira 2011, 118).

"In the early stages of the Atlantic slave trade, capture was sometimes undertaken by the European traders themselves, but by the seventeenth century, the trade was supplied directly by Africans" (Higman 2011, 493).

9. Gastrointestinal diseases, fevers, and respiratory illnesses were the most common causes of death during the voyage (Steckel and Jensen 1986, 62).

10. Manning 1990, 257. This figure is supported by data from the Slave Voyages database, which suggests that of the 12.5 million people who were loaded onto slave ships in Africa, 10.7 million disembarked alive in the Americas (Slave Voyages 2018).

11. Blackburn 2010, 17 (general), 133 (cacao, gold, mercury, and silver), 258 (rice as a plantation staple in Barbados), 397 (gold, sugar, coffee, tobacco, rice, cotton, indigo, pimientos, dried meat, and more as slave-produced exports from Brazil).

12. Blackburn 2010, 331–334. Eighteen-hour workdays are mentioned in Blackburn (2010, 260; 1997, 260). Regular days of at least ten hours are also mentioned in Blackburn (2010, 339, 424).

13. Blackburn 2010. The figure of twenty years is for Trinidad (John 1988). Records from one South Carolina rice plantation between 1800 and 1849 also indicate a life expectancy at birth of about twenty (McCandless 2011, 129).

14. Stampp 1956; as quoted in Gutman 1975, 36.

15. "The continued currency of ideas supportive of slavery was to combine the notion that particular traits—seen as flaws of origin or defects of civilisation—justified enslavement and the idea that developed chattel slavery was itself a sign of civilisation" (Blackburn 2010, 63). It should be noted that North American slaveholders did actively lobby for various legal changes because the English law on which the colonies' legal systems were based lacked some rules needed to sustain and protect their business. These included measures that prevented enslaved people from converting to Christianity in order to be set free (Walsh 2011, 413).

16. Plato does not directly address the morality or immorality of slavery, but in *Laws* he seems to condone slavery, suggesting that by virtue of their status enslaved people should receive stricter punishment: "Slaves ought to be punished as they deserve, and not admonished as if they were freemen, which will only make them conceited" (Plato 2010, 293).

In *Politics*, Aristotle writes, "For that some should rule and others be ruled is a thing not only necessary, but expedient; from the hour of their birth, some are marked out for subjection, others for rule" (Aristotle 1885, 7), and "It is manifest therefore that there are cases of people of whom some are freemen and the others slaves by nature, and for these slavery is an institution both expedient and just" (Aristotle 1932, 23–25).

"To mention just one example, in Surinam one uses red slaves (Americans) only for domestic work, because they are too weak for work in the field. For field work one needs negroes" (Kant 1912, 438; quoted in Kleingeld 2007, 576). "Americans and Negroes cannot govern themselves. Thus, [they] serve only as slaves" (Kant 1913, 878; as quoted in Kleingeld 2007, 577).

17. For example, the Haitian Revolution of 1791, the 1823 Demerara Rebellion, and the 1831 Jamaican Christmas Rebellion all played important roles in advancing the abolitionist cause in Great Britain. Michael Taylor (2021, 22) wrote that "the Demerara Rebellion of 1823 was a critical milestone in the history and downfall of slavery in the British Empire." Historian Franklin W. Knight (2000, 114) wrote that the revolution in Haiti "cast an inevitable shadow over all slave societies. Antislavery movements grew stronger and bolder, especially in Great Britain." Somewhat similarly, the influence of the Jamaican rebellion, which "convinced many Britons . . . that the endurance of slavery risked repeated scenes of bloodshed," is discussed by Taylor (2021, 191).

18. Brown 2006, 30.

19. The key figures included Peter Cornelius Plockhoy, a Mennonite; Francis Daniel Pastorius, a Lutheran; and the Quakers William Edmundson, George Keith, John Hepburn, and Ralph Sandiford. George Fox, the founder of Quakerism, had earlier made some timid antislavery comments, recommending that enslaved people be freed "after a considerable Term of Years, if they have served faithfully" (Fox 1676, 16), but he never came close to recommending abolition, and he was more concerned about the corrosive impact of slavery on slave owners than the suffering of the enslaved people themselves.

My principal source on Lay's life is Marcus Rediker's *The Fearless Benjamin Lay* (2017). Of the other early antislavery activists, Plockhoy seems to have been the first. He was a Mennonite who founded a settlement on the Delaware Bay in 1663 where slavery was not allowed, but by 1664 he was in Germantown, just north of Philadelphia. It is striking that it was in Germantown in 1688 that Mennonite converts to Quakerism like Pastorius issued an antislavery petition.

20. Rediker 2017a, 2017b.

21. Rediker 2017a, Chapter 5, Introduction.

22. Rediker 2017a, Chapters 5–6.

23. "Exhausted, emaciated workers staggered into their waterfront shop, buying, begging, and sometimes stealing small items and food. Early on, Benjamin responded to the theft in anger, lashing a few of the culprits, but he soon understood that this monstrous slave society called Barbados had been built by bigger thieves, who sought not subsistence but riches. Wracked with guilt for having behaved like a slave master, Benjamin decided to educate himself by talking with the enslaved and learning about their lives" (Rediker 2017a, 47).

24. Rediker 2017a, Chapter 2.

25. This is Rediker's (2017a, 83) account of Lydia Childs's account of a story told to her by Isaac Hopper, a nineteenth-century Quaker abolitionist who followed in Lay's footsteps, which Hopper says he had heard as a child.

26. Rediker 2017a, Chapter 4.

27. Rediker 2017a, Conclusion.

28. Vaux 1815.

29. "Woolman was in all likelihood present for the bladder-of-blood spectacle that took place in Burlington, New Jersey" (Rediker 2017a, 187).

30. Rush 1891.

31. Rediker 2017a, Chapter 3.

32. Quoted in Cole 1968, 43.

33. "If there was an eighteenth-century abolitionist who matched the pivotal role of William Lloyd Garrison in the nineteenth century, it was Anthony Benezet. . . . Benezet occupies a pride of place in early abolitionist thought, as his ideas transcended the boundaries of Quakerism" (Sinha 2016, 20–22).

34. These figures come from Soderlund (1995, 34). Note that we can only measure the decline in slave owning among Quakers for whom records exist, which may not be a representative sample of all Quakers at the time. It seems likely, though, that this group is sufficiently representative that we can infer a general decline in slave owning among Quakers, especially given the size of the decline.

35. Rediker 2017a, Chapter 6.

36. Drake 1950, 46.

37. James Oglethorpe, for example, the founder of the colony of Georgia in 1733, had the trustees of the colony expressly forbid slavery there because he worried that it would make its White colonists lazy and cruel. Only later, after becoming close friends with Granville Sharp, did Oglethorpe become involved with the abolitionist movement. Among the early moralists who condemned slavery, Samuel Sewall in 1700 made the argument that

the institution corrupted the slave owners because they were tempted to rape the enslaved people they oppressed.

38. A papal bull of 1537, for example, forbade the enslavement of Indigenous people living in the Americas because Jesus said all people could be converted, making them worthy of basic, humane treatment. However, the bull was evidently ignored. See Sinha (2016, 10) for an overview of sixteenth-century condemnations of slavery by Catholic clerics.

Bartolomé de las Casas, who lived in the sixteenth century, is often mentioned as an example of someone opposed to slavery. Having been horrified by the massacre and enslavement of Indigenous peoples by Spanish colonists in the Americas, he at first recommended replacing them with enslaved people from Africa, apparently in the belief that they had been enslaved for "just" reasons, such as their being convicts or captives in just wars. He later regretted this recommendation after he learned that many enslaved Africans had been kidnapped, their families torn apart, because of raids and unjust wars of conquest. His opposition thus originally stemmed from his view that some people were unjustly enslaved and from his disapproval of the cruelty that ensued on plantations, rather than from a condemnation of slavery as an institution. In theory, at least, he conceded that slavery arising from a just war could be legitimate (Pennington 2018, 111).

George Fox, the founder of the Society of Friends, is an example of those who argued for releasing enslaved people as a matter of charity. In 1657 he called on Quakers to be merciful to their slaves. He later published a short book in 1676 based on speeches he gave in Barbados. He suggested that it would be "very acceptable to the Lord" if masters freed their slaves "after a considerable Term of Years, if they have served faithfully" (Fox 1676, 16).

39. See, for example, the works of Francis Hutcheson or Denis Diderot.

40. See, for example, the abolition of slavery in China in AD 17 by a usurping minister, Wang Mang, who wished to limit the power of landowning families. Or see the sixteenth-century manumissions by Mughal emperor Akbar, who appears to have been concerned that the export of enslaved Indians was causing population decline, that enslavement was reducing the number of taxpaying peasants, and that military officers were building up independent power bases by transforming enslaved people into personal retainers or enriching themselves by selling them (Eaton 2006, 11–12). The widespread reduction of various unfreedoms in 1723–1730 by China's Yongzheng Emperor appears to have been due to a similar concern about the power of the nobility, in that he hoped to create an undifferentiated class of free subjects under his direct rule (Crossley 2011).

41. Hochschild (2005, 5; emphasis in original) goes further than this, suggesting that the British abolitionist campaign was

> something never seen before: it was the first time a large number of people became outraged, and stayed outraged for many years, over someone *else's* rights. And most startling of all, the rights of people of another color, on another continent. No one was more taken aback by this than Stephen Fuller, the London agent for Jamaica's planters, an absentee plantation owner himself and a central figure in the proslavery lobby. As tens of thousands of protesters signed petitions to Parliament, Fuller was amazed that these were "stating no grievance or injury of any kind or sort, affecting the Petitioners themselves." His bafflement is understandable. He was seeing something new in history.

42. Hornick 1975. I'm deliberately capitalizing both "Black" and "White" when referring to racial or cultural groups or concepts, following the recommendation of, e.g., the National Association of Black Journalists (2020) and the Diversity Style Guide (Kanigel 2022). Note that especially the capitalization of "White" is a matter of debate, with, for instance, the Associated Press (Bauder 2020) and the *New York Times* (Coleman 2020) capitalizing "Black" but not "white."

43. Hornick 1975.
44. Brendlinger 1997, 121–122.
45. Hanley 2019, 180.
46. UK Parliament 2021.
47. Sullivan 2020.
48. C. L. Brown 2007, 292.
49. Our World in Data 2021c.
50. Gershoff 2017.
51. On the scale of international migration, see UN (2019a).
52. Pritchett 2018, 4.
53. For the number of land animals raised and killed in factory farms, see FAO (2021) and Anthis and Reese Anthis (2019). If we include farmed fish, the number of animals in factory farms could rise to over a trillion (Mood and Brooke 2019).
54. ScotsCare, n.d.
55. This is according to the UK's Office for National Statistics (2018). By some measures, though, Edinburgh's gross domestic product per capita is actually higher than London's (Istrate and Nadeau 2012).
56. Gould 1989.
57. T. Y. W. Wong 2019.
58. Losos 2017, Conclusion, Chapter 3.
59. Martini et al. 2021; Blount et al. 2018.
60. Some popular claims about specific instances of carcinisation are, however, of dubious veracity. McLaughlin and Lemaitre (1997, 117) conclude that "carcinization, if meaning only acquisition of a crab-like body form, must be acknowledged as a fact. However, . . . the evolution of a crab-like body form from a shell-dwelling pagurid is, in our opinion fictitious, not factual."
61. Van Cleve and Weissman 2015.
62. De Robertis 2008.
63. The theory of cultural evolution has increasingly been a focus of serious academic study over the past four decades, in particular since the publication of Robert Boyd and Peter Richerson's *Culture and the Evolutionary Process* (1988), which showed how mathematical models from evolutionary biology could be applied to cultural change. We should be careful to distinguish this theory from the related field of memetics, which is of more dubious scientific standing (Chvaja 2020).
64. Bowles and Gintis 2011; Henrich 2004.
65. Henrich 2018, Chapter 10.
66. Curry et al. 2019.
67. It turns out to be surprisingly hard to get good data on the proportion of vegetarians in different countries around the world. As an example of the problems surveys of vegetarianism face, one large study found that about 40 percent of self-identified vegetarians consumed meat or poultry products (Juan et al. 2015). What's more, different estimates of the proportion of vegetarians in a given country usually vary quite a lot. The numbers I've used here are from a global survey that relied on self-reported dietary habits, so I expect they significantly overestimate the actual prevalence of vegetarianism (Nielsen 2016, 8). Still, the differences between regions are more important than the absolute proportions, and I don't expect those would disappear even if we were able to adjust for unreliable self-reporting.
68. OECD 2021a.
69. Tatz and Higgins 2016, 214; Martin 2014, Appendix I. In addition to the Albigensian Crusade, oppressive policies instituted by the French king Louis IX contributed to Catharism's extermination (*Encyclopedia Britannica* 2007).
70. Jonsen and Toulmin 1989, 203.

71. Ellman 2002, 1162.
72. Becker 1998, 176.
73. Short 2005, Chapter 11.
74. Locard 2005.
75. *New York Times* 2018.
76. Theodorou and Sandstrom 2015.
77. The proportion of the population saying men have more right to a job is from the World Values Survey, Wave 6 (Inglehart et al. 2014). Workforce participation rate from International Labour Organization estimates are via Our World in Data (2021b).
78. Funk et al. 2020; note that China was excluded from this survey. To check the result that India has unusually positive attitudes to human genetic enhancement, I asked psychologists Lucius Caviola and David Althaus to try to replicate this result, surveying 164 Indians and 167 people from the United States. The same effect was found, although it wasn't as strong: 49 percent of Indians thought that it was appropriate to use technology to change a baby's genetic characteristics to make the baby more intelligent; only 14 percent of US respondents did.
79. Although there have been many surveys on attitudes towards genetic enhancement (a recent systematic review included forty-one studies), it's difficult to find reliable, comparable data for multiple countries (i.e., large studies that asked people in multiple countries the same question). This is important because it seems likely that questions about such a controversial, technical subject are vulnerable to respondent misunderstanding and framing effects. Still, it's telling that a Pew Research survey found that support for nontherapeutic genetic enhancement did not exceed 20 percent in any North American or European country, while support across Asia was much more variable and higher on average. The bioethicist Darryl Macer writes that researchers have generally found higher support for genetic screening and gene therapy practices among respondents in China, India, and Thailand than in other Asian countries (Macer 2012). However, survey data on public opinion in China, in particular, is noisy and far from conclusive (see, e.g., Zhang and Lie 2018).
80. Inglehart et al. 2014; UN 2019a. Again, data on rates of vegetarianism do not seem that reliable. The ten-to-one ratio between India and Brazil comes from a study that estimated vegetarianism prevalence using data from household consumption surveys, which strikes me as more reliable than the typical self-reported data. That study estimated that 3.6 percent of Brazilians are vegetarian, while 34 percent of Indians are (Leahy et al. 2010, 23, Table A2). A caveat here is that this paper used old data: the data for Brazil are from 1997 and the data for India are from 1998. Other estimates vary, and some show a smaller difference between Brazil and India. More at whatweowethefuture.com/notes.
81. Gallup 2018. Sri Lanka was not included in the survey in 2017, but it ranked as one of the top ten countries in the World Giving Index each year from 2013 to 2016 and was ranked twenty-seventh in 2018. Myanmar was in the top ten each year from 2013 to 2018 (Charities Aid Foundation 2019).
82. More precisely, I think it's more likely than not that in somewhere between ten and ninety of those reruns, at the point at which the world has today's level of technological development, at least 1 percent of the world population would be enslaved.
83. Brown 2007, 289. By "the economic interpretation," Brown is referring to Williams's account of the abolition of the slave trade in 1807, which Brown describes as follows:

> Two changes in the economic climate during the Age of Revolutions were crucial to Williams. There was, first, the separation of the North American colonies from the Caribbean plantations and a consequent decline in the British commitment to the West Indian monopoly on the home market. In addition to the rise of free-trade ideology there was, secondly, Williams argued, a crisis of overproduction in the West

> Indian colonies in 1806 and 1807 that made the abolition of the British slave trade feasible. Williams acknowledged the determination and skill of the abolitionist leadership, but insisted that they prevailed only because the economic interests of the nation had shifted dramatically by the early nineteenth century. (Brown 2007, 289)

84. Michael Taylor (personal correspondence, November 15, 2021) was willing to endorse this slightly distinct claim: "Since the publication of *Econocide*, ever fewer historians of slavery have maintained an explicitly economic interpretation of British abolition." Adam Hochschild (personal communication, November 6, 2021) wished to emphasise his belief that Williams still deserves much credit for pointing out how the profits produced by slave labour in the British West Indies helped fund the start of Britain's Industrial Revolution.

Though David Brion Davis has sadly passed away, it's clear that he would also have endorsed this view of the economic interpretation. He summarized Williams's argument as "The British abolished the slave trade and slavery for purely economic reasons" and said that "this decline thesis is anything but 'alive and well.' It has been undermined by a vast mountain of empirical evidence and has been repudiated by the world's leading authorities on New World slavery, the transatlantic slave trade, and the British abolition movement" (D. B. Davis and Solow 2012). He referenced—along with Seymour Drescher—David Eltis, David Richardson, Barry Higman, John J. McCusker, J. R. Ward, and Robin Blackburn as eminent scholars who reject Williams's thesis concerning the cause of British abolition.

85. According to Kaufmann and Pape (1999, 634), British colonies produced 55 percent of the world's sugar in 1805–1806, representing about 4 percent of the country's national income. In the late eighteenth and early nineteenth centuries, Britain, with a population 10 percent the size of continental Europe's, consumed 80 percent as much sugar as the continental countries combined. From Drescher's *Econocide*:

> The most interesting information about the sugar market from 1787 to 1806, however, is not in the aggregate figures for the North Atlantic. There was a dramatic shift in consumption patterns between Britain and the rest of Europe. Between 1787 and 1805–1806 the British increased their consumption of sugar by over one-third. They also increased their share of North Atlantic imports from 27 to 39 percent. During this same period, continental Europe's purchases of sugar dropped by more than one-fifth, while its share of North Atlantic imports decreased from almost two-thirds to just one-half (see table 25). In other words, Britain, with less than one-tenth of the population of the Continent, was consuming four-fifths as much sugar as the mainland in 1805–1806. (Drescher 2010, 126)

86. The effect of the Act of Emancipation was not to lower the price of sugar to the British public, but to raise it. The increased price was due partly to higher sugar duties which were used to help finance the compensation of the planters. The main reason for the rise in sugar prices, however, was the fall in the productivity of the West Indian plantations. Not only did labor discipline on the sugar estates decline, but once free, the ex-slaves fled these estates in droves, moving onto vacant land where they produced foodstuffs (either for self-subsistence or for sale in the local markets) instead of sugar. West Indian exports of sugar declined and the price of sugar rose sharply in Britain. British consumers paid 48 percent more for sugar during the first four years of freedom than they had to pay during the last four years of slavery. Indeed, between 1835 and 1842 the extra cost of sugar to the British was about £21 million, thus raising the British outlay for emancipation to over £40 million. No wonder Cobbett and other radical leaders were so hostile to the antislavery campaign. Distributed to the urban poor, that sum could have doubled their income for a decade. (Fogel 1994, 229)

87. Slave Voyages 2018.

88. "It was necessary to obtain a bill that would satisfy both the abolitionists and the West Indian lobby since Wellington had let it be known that the Lords would block any bill 'which the West Indians, as an important interest group, would not accept.' . . . Under the Emancipation Act, the planters were to be compensated for the loss of their property. About half of the compensation would be in the form of a cash payment (£20 million) to the planters at the direct expense of British taxpayers" (Fogel 1994, 228).

89. Chantrill 2021.

90. Fogel 1994.

91. As quoted in Brown 2007, 291. The 2 percent estimate is from Pape and Kaufman (1999).

92. We can also simply study the particular cases of these treaties:

> Between 1807 and 1823 Wilberforce and other abolitionist leaders generally preferred to rely on their personal influence with cabinet members rather than on public campaigns. The one major exception took place in 1814 when Viscount Castlereagh seemed ready to let France resume the slave trade in order to win other concessions from Louis XVIII at the Congress of Vienna. On short notice the abolitionists launched a nationwide petition campaign to press for articles against the trade at the peace negotiations. In a little over a month some 800 petitions with about 750,000 names were gathered. It was a public campaign of unprecedented magnitude. About one out of every eight adults had aligned themselves with the demand for international agreements to end the slave trade. Although "irritated by this abolitionist pressure," Castlereagh felt "compelled" to make the slave trade an issue and "to use both threats and bribes" to obtain an agreement. (Fogel 1994, 217–218)

93. Burrows and Shlomowitz 1992.

94. A full list of the sectors in which enslaved people are documented to have worked in ancient Greece would include agriculture, animal husbandry, metalwork, carpentry, leatherworking, weaving, mining, quarrying, housekeeping, cooking, baking, childcare, policing, commerce, business management, banking, and prostitution (Forsdyke 2021).

95. That there has recently only been a single trend in moral values is discussed in Alexander (2015), from which I got the neckties example.

96. This view is given by, for example, philosopher Michael Huemer (2016).

97. Estimates of the number of forced labourers used by the Nazis in World War II vary, but the best estimate is eleven million (Barenberg 2017). Most sources agree that about 75 percent were civilians (Davies 2006).

98. Barenberg 2017, 653.

99. Gillingham 2014.

100. It's worth noting that if the plot to which a serf was bound was sold, the serf would typically be "transferred" to the new owner along with the land (Walvin 1983).

101. The Black Death caused labour shortages that, in conjunction with growing central government power and peasant uprisings, contributed to the replacement of serfdom with a system of free peasantry by the end of the fifteenth century (*Encyclopedia Britannica* 2019b).

102. For example, Perry et al. (2021) write that between the fall of the Roman Empire and the rise of the transatlantic slave trade, "slavery continued to flourish in all parts of the world for which records and material objects have survived. In short, both the dismemberment of the Roman Empire and Columbian contact had large effects on who was enslaved but quite possibly not on the incidence of the institution across the globe" (Perry et al. 2021, 1).

103. Kahan 1973.

104. Han dynasty slavery: Wilbur (2011). Evidence for earlier slavery in China is less conclusive—see Hallett (2007) and Rodriguez (1997) for the Shang dynasty, Yates (2001)

for the Qin dynasty immediately preceding the Han, and Pulleybank (1958) for the Warring States period.

105. Eras during which reform or abolition was attempted include the Han dynasty, the Red Eyebrows rebellion, the Song dynasty, and the Ming dynasty (as discussed in Hallet [2007]).

106. "The Qing not only conquered Liaodong province and absorbed its populations of Chinese-speaking farmers, merchants, and soldiers for its own use, but it increased its campaigns for the extraction of more forced labor from Korea and China. According to the most noted scholar of Qing slavery, Wei Qingyuan, soon after the second khan's accession to the throne in 1626, the population registers enumerated more than two million domestic and agricultural slaves, compared to a probable common population of fewer than six million" (Crossley 2011, 201).

107. Hallet 2007.

108. See, e.g., Eltis 1999, 281–284.

109. Sala-Molins 2006. The National Constituent Assembly banned slavery by decree in 1794, and abolition was implemented in Saint-Domingue, Guadeloupe, and Guyana but not in Martinique, Senegal, Réunion, Mauritius, or French India (Peabody 2014).

110. Indeed, Daniel Resnick (1972) refers to Clarkson's London Society for the Abolition of the Slave Trade as the "parent" or "patron" organization of Brissot's Société.

111. Peabody 2014.

112. Fogel 1994, 9–13.

113. Sinha 2016, 35.

114. See Chapter 9 for more discussion.

115. The European Convention on Human Rights and the United Nations' "Standard Minimum Rules for the Treatment of Prisoners" both prohibit corporal punishment.

116. More precisely, 1.86 million men were drafted during the Vietnam War (US Selective Service System 2021).

117. Cook 2017, 1.

118. Cook 2017.

119. "In 1913 the trustees reported an incredible profit of nearly $937,000 for the past biennium" (W. B. Taylor 1999, 41).

120. Indeed, the prison operates to this day (Cook 2017).

121. It is difficult to say exactly how many prisoners work or how much they earn on average. The public corporation which organizes prison labour at the federal level is known as UNICOR, or Federal Prison Industries. It reports that over twenty thousand inmates, or about 8 percent of the total prisoner population, participate in its work programmes annually (US Federal Bureau of Prisons, n.d.-b). UNICOR also notes that "typical hourly pay" is between $0.23 and $1.15 per hour (US Federal Bureau of Prisons, n.d.-a). However, there are also state-level work programmes. In 2017, the *Economist* reported that the total number of prison labourers in the United States was sixty-one thousand (*Economist* 2017). However, the last full census of prisoners, which took place in 2005, reported that "about half" of all prisoners had work assignments (Stephan 2008). Since the prison population at that time was over 1.4 million, if that proportion holds today, the total number of prison labourers could be an order of magnitude higher than the *Economist*'s estimate.

122. US National Archives 2016.

123. Brown 2012, 30. In conversation, Brown took back his use of the term "accident": there were, of course, many causes of abolition; it wasn't a random event. For context, two other relevant quotes from *Moral Capital* are these: "The British abolition movement that began in the 1780s did not follow inevitably from enlightened sensibilities, social change, or a shift in economic interests" (Brown 2012, 1) and "Too often, the British campaigns of the late eighteenth century have been presented as the predictable outcome of the era, as

the logical result of cultural trends, social change, political shifts, or economic forces, as a consequence of human progress. Yet the story of how the British antislavery movement began suggests more strongly that the campaign itself was fortuitous, that it need not have developed when it did, as it did, and with the popularity that it acquired. In the end, what is remarkable about abolitionism in Britain is not that it took so long to emerge, that it was politically ineffective for many years, or that it was limited in its ambition and selective in its scope. Such movements often are. What is truly surprising about British abolitionism is that such a campaign ever should have developed at all" (461f).

> 124. Who in a position of authority, and how many in the political nation, would have elected to alienate the British planter class just years after a war for independence had been narrowly averted? This planter interest would have found it difficult to seek independence, to be sure, though one can imagine southern and Caribbean slaveholders entertaining the possibility of an alliance with a European rival, as some Saint Domingue planters did during the early years of the Haitian Revolution. Undoubtedly, southern and Caribbean propagandists would have tried to recruit northern assistance by portraying the challenge to the slaving interest as a threat to the rights of all the American colonies, both those with slaves as well as those with none. Under these circumstances, an attack on slaveholders or slave traders might have seemed needlessly provocative and dangerously divisive to those in Britain and North America sympathetic to antislavery impulses but wary of precipitating a renewed debate over taxation and representation, imperial sovereignty, and the rights of colonies. (Brown 2012, 455)

125. "Abolitionism did not confer opportunities, status, or further benefits to its proponents in France. After 1788, in fact, its association with British reform briefly tainted antislavery activism. . . . The new association of abolitionism with Jacobinism would mean that antislavery would be linked with turmoil and violence in France and Haiti after the restoration of the French monarchy. French abolitionists in the first half of the nineteenth century would have to contend not only with the proslavery interest but also with the negative associations that antislavery had acquired after Haitian independence" (Brown 2012, 459).

126. See Brown (2012, 454–462) for a full picture of a counterfactual history where a strong plantation lobby, united across Britain and its colonies, successfully fought off abolitionist pressure.

127. Taylor 2021, 13. Taylor confirmed his timeline of decades to me in correspondence. A further quote:

> The Abolition Act was neither the inevitable bequest of sweeping anti-slavery sentiment and the triumphant march of British "justice," nor was it a simple coda to the better-known campaign against the slave trade. In reality, the passage of the Act had relied upon several factors: the political collapse of the Tories which led to Reform and the return of a sympathetic House of Commons; the persistent pressure applied by the Anti-Slavery and Agency societies; and the violent slave resistance that finally convinced the British public of the immoral, unsustainable nature of slavery. Until those factors combined in the early 1830s, defending slavery was a tenable, popular position for British conservatives, imperialists, economists, and more besides. Until 1833, slavery had been an essential part of British national life, as much as the Church of England, the monarchy, or the liberties granted by the Glorious Revolution. (Taylor 2021, 205–206)

128. The parliamentary reforms included the Catholic Relief Act of 1829, which completed Catholic emancipation and sowed discord among conservative parliamentarians, and the Reform Act of 1932, which broadened the electorate (Taylor 2021).

129. Taylor, personal communication (September 28, 2021).

130. Taylor 2021, 100.

131. Estimates for the population of Ethiopia and the number of enslaved people are highly uncertain, but the majority of sources agree with these figures (see Coleman 2008, 73n34).

132. Goitom 2012.

133. Klein 2014, xxiv.

134. While, again, reliable estimates of the enslaved population at the time do not exist, the British committees established to look into the issue and push for Saudi abolition reportedly believed there were "between 15,000 and 30,000 slaves" in the country at the time (Miers 2005, 119).

135. Klein 2014, xxiv–xxv.

136. Kline 2010; G. R. Searle 1979; Björkman and Widmalm 2010.

137. Cahill 2013.

138. Rush 1891.

139. Cotra (2017) provides a detailed discussion of whether hens are better off in cage-free housing than in battery cages. Šimčikas (2019) estimates the number of hens affected by corporations' cage-free pledges.

140. Garcés has written about this at length in *Grilled: Turning Adversaries into Allies to Change the Chicken Industry* (2019).

Chapter 4: Value Lock-In

1. This is a bit of a misnomer, however, since very few of the philosophies of the time were developed into formal schools with students systematically attempting to study and expand their doctrines—arguably only the Mohists and the Confucians had this status. Moreover, there was considerable overlap and interchange between schools, especially in later periods. Regarding the dating, note that the erosion of Zhou authority was gradual and that the beginning of the "Hundred Schools of Thought" period is often given as the sixth century BC, towards the end of the Spring and Autumn period and prior to the fifth century BC start dates most commonly given for the Warring States period. More at whatweowethefuture.com/notes.

2. Fang 2014.

3. Some scholars also add Yin-yang and the School of Names, bringing the total number of schools to six.

4. One robust account of the concept of "sageliness" in Chinese philosophy can be found in Feng (1997, 6–9).

5. D. Wong 2021.

6. Csikszentmihalyi 2020.

7. Note that Legalists were not a self-aware and organized intellectual current; rather, the name was coined as a post-factum categorization of certain thinkers and texts. See other concerns with this naming convention in Goldin (2011).

8. Lao Tzu 2003. For modern views on the history of Daoism, see whatweowethefuture.com/notes.

9. Mengzi 3B9.9; quoted in Van Norden 2007, 185.

10. The Mòzǐ principle of *jiān ài* is sometimes translated as "universal love" (Van Norden 2019).

11. These principles show up in the ten Mohist doctrines, namely "moderation in use" (Fraser 2020).

12. Gladstone 2015.

13. The Xúnzǐ quote is from Eric L. Hutton's (2005, 264) translation.

14. More precisely, the Qin were influenced by thinkers such as Shang Yang, Shen Buhai, and Han Fei, who would only later be called Legalists (Pines 2018, Section 1).

15. Nylan 2001, 23.

16. It is often claimed that the scholars were buried alive; however, according to sinologist Derk Bodde, the relevant term in the Chinese original simply means "slain." More broadly, the historiography of the Qin biblioclasm is rife with myths. Its most popular account is from Han dynasty scholar Sima Qian, whom modern historians consider unreliable because he was incentivised to disparage the Qin. My account follows the modern consensus, which agrees that books were burned and scholars executed (Kramers 1986, Chapters 1, 14).

17. It is commonly claimed that the influence of Confucian thought was wholly eradicated. This is again due to the account by Han scholar Sima Qian, which modern historians consider exaggerated.

18. Tanner 2009, 87; C. C. Müller, 2021; Bodde 1986, 78–81.

19. *Encyclopedia Britannica* 2019d, 2021e.

20. Csikszentmihalyi 2006; Kramers 1986.

21. Goldin 2011, 99–100. According to a first century BC Chinese history text, one Han Confucian, Master Yuan Gu, was even locked in a pigpen and forced to fight a boar because he had bluntly told the empress dowager that the *Daodejing* (a classic Daoist text) was "the saying of a menial, nothing more!" (Sima 1971, 364).

22. This account of Confucianism's rise follows Liang Cai's *Witchcraft and the Rise of the First Confucian Empire* (2014). Cai rejects the common view, elaborated in the 1930s by Homer Dubs, that Confucianism became state doctrine under the earlier Han emperor Wudi. A quantitative analysis shows that Wudi employed only six Confucian officials in his half-century reign, while twelve achieved leading positions under Xuan's twenty-five-year rule (Cai 2014, 29). Cai (3) argues that "to legitimate their success," these Confucians "read it back into history, retrospectively constructing a flourishing Confucian community under Emperor Wu." For the common view, see Dubs (1938).

23. Kohn 2000.

24. Morris 2010, Chapter 7.

25. To be sure, Buddhism and Daoism still had their place in the private religious life of citizens, but Confucianism was the philosophy of public life and of government. For San Zi Jing, see Zhu and Hu (2011).

26. In the last chapter, I talked about value changes as being unusually predictable in their impact. Remarkably, the idea of the predictability of moral influence seems to have been understood by Confucius himself. *The Analects* (that is, the sayings of Confucius) contains the following passage:

> Zizhang asked, "Can the future be known even at a remove of ten generations?" Confucius replied, "The Yin house was founded on the ceremonial traditions [Li] of the Xia, its predecessor, and amended them in ways known to us. Our own Zhou house was founded on the ceremonial traditions of the Yin, its predecessor, amended in ways known to us. And should some other house filially succeed our Zhou, the future can still be known even at a remove of one hundred generations." (Confucius 2020, 38)

27. This is according to the World Values Survey, a global survey conducted in over a hundred countries every five years. The idea that distinctive cultural histories shape differences in the typical responses from people from different nations comes from the survey's own "World Cultural Map," which uses factor analysis to map countries along two dimensions: traditional vs. secular values and survival vs. self-expression values. A distinct cluster of "Confucian heritage" countries like China and South Korea score highly on secular values while scoring about average on survival vs. self-expression. In contrast, "Protestant" European countries are much higher on self-expression, while "Orthodox" European countries score higher on survival values (World Values Survey 7 2020, The Inglehart-Welzel World Cultural Map).

This analysis needs a couple of caveats. First, the data used for the World Cultural Map reflect "only a handful" of the beliefs and values covered by the World Values Survey. One could question whether the specific indicators used to build out the "traditional values" factor, for example, accurately reflect the meaning of that term as we typically understand it. Second, conducting a high-powered study across so many countries is an inherently challenging endeavour. Sometimes the average response on a given question in a given country changes quite dramatically from one survey to the next. This is to be expected because of statistical variation, but it does mean that one should not take the results of one edition of the survey to be definitive. For these reasons, I think the results of the World Values Survey, as well as the World Cultural Map, are suggestive but not conclusive evidence of enduring cultural differences across countries.

28. The body of academic work known as persistence studies is highly relevant to the persistence of values (for a review, see Cioni et al. [2020]). In a previous draft of this book, I discussed some striking claims advanced in that literature, including about longterm harms from slavery (Nunn 2008; Nunn and Wantchekon 2011). However, prompted by criticisms of the methodology employed in persistence studies (Kelly 2019, 2020; Arroyo, Abad, and Maurer 2021), I commissioned a quantitative review of some key papers (Sevilla 2021ab, available on the book's website). As a result, I did not feel confident enough in the persistence studies findings to include them in this book. For responses to recent criticism by a proponent of persistence studies, see Voth (2021).

29. There are no records of all global book sales, so global sales figures are uncertain. According to the *Guinness World Records* website, five to seven billion copies of the Bible have been printed in total as of 2021 (Guinness World Records, 2021). The *Economist* claims that a hundred million Bibles are sold or given away by churches every year (*Economist* 2007). For comparison, between 1997 and 2018, the Harry Potter series sold five hundred million copies (Eyre 2018; Griese 2010).

30. Estimating sales of the Quran is as difficult as estimating sales of the Bible. The *Southern Review of Books* has "guesstimated" that the Quran has sold eight hundred million copies (Griese 2010). Because the Muslim population is increasing over time, sales are likely also increasing. The nearest competitor is Mao Zedong's *Little Red Book*, with eight hundred to nine hundred million sales, though demand for that has declined substantially since Deng Xiaoping's reforms in the 1970s (Griese 2010). According to *Foreign Policy*, in 2013, the *Little Red Book* was out of print in China (Fish 2013).

31. China Global Television Network 2017.

32. Babylonian Talmud Yevamot 69b as quoted in Schenker 2008, 271; Catholic News Agency 2017; Crane 2014; Prainsack 2006.

33. Kadam and Deshmukh 2020.

34. For a parallel discussion of value lock-in as a type of "existential catastrophe," see Ord (2020, 157).

35. For more detail on how artificial intelligence might enable value lock-in or otherwise allow contingent features of civilisation to persist for a very long time, see Finnveden, Riedel, and Shulman (2022).

36. Silver et al. 2016, 2017. DeepMind claims that AlphaGo "was a decade ahead of its time" (DeepMind 2020). This might refer to a 2014 prediction by Rémi Coulom, the developer of one of the best Go programmes prior to AlphaGo (Levinovitz 2014). However, this may be exaggerated. Go programmes had been reliably improving for years, and a simple trend extrapolation would have predicted that programmes would beat the best human players within a few years of 2016—see, e.g., Katja Grace (2013, Section 5.2). After correcting for the unprecedented amount of hardware DeepMind was willing to employ, it is not clear whether AlphaGo deviates from the trend of algorithmic improvements at all (Brundage 2016).

37. More specifically, most AI breakthroughs have been due to a particular approach to machine learning that uses multilayered neural networks, known as "deep learning" (Goodfellow et al. 2016; LeCun et al. 2015). At the time of writing, the state-of-the-art AI for text-based applications are so-called transformers, which include Google's BERT and OpenAI's GPT-3 (T. Brown et al. 2020; Devlin et al. 2019; Vaswani et al. 2017). Transformers have also been successfully used for tasks involving audio (Child et al. 2019), images (M. Chen et al. 2020; Dosovitskiy et al. 2021), and video (Wang et al. 2021). The highest-profile AI achievements in real-time strategy games were DeepMind's AlphaStar defeat of human grandmasters in the game *StarCraft II* and the OpenAI Five's defeat of human world champions in *Dota 2* (OpenAI et al. 2019; Vinyals et al. 2019). Early successes in image classification (see, e.g., Krizhevsky et al. 2012) are widely seen as having been key for demonstrating the potential of deep learning. See also the following: speech recognition, Abdel-Hamid et al. (2014); Ravanelli et al. (2019); music, Briot et al. (2020); Choi et al. (2018); *Magenta* (n.d.); visual art, Gatys et al. (2016); Lecoutre et al. (2017). Building on astonishing progress demonstrated by Ramesh et al. (2021), the ability to create images from text descriptions by combining two AI systems known as VQGAN (Esser et al. 2021) and CLIP (OpenAI 2021b; Radford et al. 2021) caused a Twitter sensation (Miranda 2021).

38. "BERT is now used in every English search, Google says, and it's deployed across a range of languages, including Spanish, Portuguese, Hindi, Arabic, and German" (Wiggers 2020). BERT is an example of a transformer (see the previous endnote).

39. See whatweowethefuture.com/notes.

40. Discussions about potential large-scale impacts from future AI systems suffer from a proliferation of terminology: apart from AGI, people have talked about transformative AI (Cotra 2020; Karnofsky 2016), smarter-than-human AI (Machine Intelligence Research Institute, n.d.), superintelligence (Bostrom 1998, 2014a), ultraintelligent machines (Good 1966), advanced AI (Center for the Governance of AI, n.d.), high-level machine intelligence (Grace et al. 2018; and, using a slightly different definition, V. C. Müller and Bostrom 2016), comprehensive AI services (Drexler 2019), strong AI (J. R. Searle 1980, but since used in a variety of different ways), and human-level AI (AI Impacts, n.d.-c). I'm using the term "AGI" simply because it is probably the most widely used one, and its definition is easy to understand. However, in this chapter, I am interested in any way in which AI could enable permanent value lock-in, and by using "AGI" as opposed to any of the other terms mentioned previously, I do not intend to exclude any possibility for *how* this could happen. For instance, perhaps value lock-in could come about through the cumulative effects of deploying multiple different AI systems rather than one AGI, or perhaps AI might enable value lock-in when still lacking some key capabilities, such as the ability to directly manipulate the physical world (if robotics lags behind other areas of AI).

41. DeepMind 2020.

42. "Our teams research and build safe AI systems. We're committed to solving intelligence, to advance science and benefit humanity" (DeepMind, n.d.). "Our mission is to ensure that artificial general intelligence benefits all of humanity" (OpenAI 2021a).

43. See whatweowethefuture.com/notes.

44. Silver et al. 2018.

45. Schrittwieser et al. 2020a, 2020b.

46. My grandmother Daphne S Crouch is listed on the Bletchley Park Roll of Honour (Bletchley Park, n.d.-a) and commemorated at brick location E1:297 in Bletchley Park's (n.d.-b) digital Codebreakers' Wall. The fact that Good worked at Bletchley Park is well known (see, e.g., *Guardian* 2009). The idea that thinking machines would at some point quickly overtake human intelligence and would then "take control, in the way that is mentioned in Samuel Butler's *Erewhon*" was raised by Turing (1951, 475), but the classic statement of the idea comes from Good (1966, 33; emphasis in original): "Let an ultraintelligent

machine be defined as a machine that can far surpass all the intellectual activities of any man however clever. Since the design of machines is one of these intellectual activities, an ultraintelligent machine could design even better machines; there would then unquestionably be an 'intelligence explosion,' and the intelligence of man would be left far behind. . . . Thus the first ultraintelligent machine is the *last* invention that man need ever make, provided that the machine is docile enough to tell us how to keep it under control."

47. Nordhaus 2021. For an overview of economists' work on the implications of AI for economic growth, see Trammell and Korinek (2020).

48. This implication of Nordhaus's model is explained in Trammell and Korinek (2020, Section 3.2).

49. This is what Nordhaus (2021, Section VI) calls a "supply-side singularity." While this is the focus of Nordhaus's paper, he also discusses two other ways through which AI could accelerate growth. More at whatweowethefuture.com/notes.

50. Callaway 2020. "This computational work represents a stunning advance on the protein-folding problem, a 50-year-old grand challenge in biology. It has occurred decades before many people in the field would have predicted. It will be exciting to see the many ways in which it will fundamentally change biological research." Professor Venki Ramakrishnan, Nobel laureate and president of the Royal Society 2015–2020, quoted in AlphaFold Team (2020).

51. Aghion et al. 2019, Section 9.4.1, examples 2–4. More generally, the arguably empirically most plausible explanation of economic growth—as captured in so-called semiendogenous growth models (for a review, see Jones [2021])—implies accelerating growth once AI systems can substitute for human labour, assuming that the population of AI workers could grow faster than the current population of humans. For an excellent exposition of this and other arguments for why AGI could plausibly cause a growth explosion, see Tom Davidson (2021b).

52. The critical questions include whether ideas (of the kind that drive productivity-enhancing technological progress) are getting easier or harder to find over time (see, e.g., Aghion et al. 2019, 251) and how easily AI can substitute for other inputs or outputs—a property that economists measure with a parameter known as "elasticity of substitution." The latter point is highlighted both by Aghion et al. (2019, 238)—"Economic growth may be constrained not by what we do well but rather by what is essential and yet hard to improve"—and Nordhaus (2021, 311): "The key parameter [for whether the model implies a supply-side singularity] is the elasticity of substitution in production."

53. For the history of global economic growth, see, for instance, DeLong (1998). For an overview of other data sources, which give similar numbers, see Roodman's (2020a) data and Roser's (2013b) sources.

54. Hanson 2000.

55. See the discussion in Garfinkel (2020).

56. Thanks to Paul Christiano for bringing these issues to my attention. (See also Christiano 2017; Roodman 2020b.)

57. Again, this consideration was noted by the early computer science pioneers: when discussing risks from AI, Turing (1951, 475) noted that "there would be no question of the machines dying."

58. *Pong* was first released in 1972 as an arcade game (*Encyclopedia Britannica* 2020d)—a bulky, coin-operated machine at which one could play nothing but *Pong* (see Winter [n.d.-b] for images and a more detailed history). However, this version did not involve any software. More at whatweowethefuture.com/notes.

59. It's available, for instance, on the RetroGames website (Atari 1977).

60. Bostrom and Sandberg 2008; Hanson 2016; Sandberg 2013.

61. See whatweowethefuture.com/notes.

62. See whatweowethefuture.com/notes.

63. *Encyclopedia Britannica* 2021b.

64. "Moreover, even reasonable normative views often recommend that they be locked in—for otherwise a tempting rival view may take over, with (allegedly) disastrous results" (Ord 2020, 157).

65. The seminal biophysicist Alfred J. Lotka (1922, 152) used "the persistence of stable forms" as synonymous with the principle of natural selection itself.

66. For Austrian poet Rainer Maria Rilke, "the epic [of Gilgamesh] was first and foremost 'das Epos der Todesfurcht,' the epic about the fear of death" (George 2003, xiii). More at whatweowethefuture.com/notes.

67. Cedzich 2001, 1.

68. Needham 1997.

69. The worry that future technology could make totalitarianism last much longer was also discussed by Caplan (2008, Section 22.3.1) and Belfield (forthcoming).

70. The source is the dissident Russian brothers Zhores and Roy Medvedev (2006, 4).

71. Based on testimony from Kim Il-sung's former personal physician Kim So-Yeon, who defected to South Korea in 1992 (Hancocks 2014).

72. *Guardian* 2012.

73. Isaak 2020.

74. Friend et al. 2017.

75. Fortson 2017.

76. Alcor 2020.

77. "Altman tells *MIT Technology Review* he's pretty sure minds will be digitized in his lifetime" (Regalado 2018).

78. Cotra 2021.

79. The argument that, for a wide range of ultimate goals, it is useful for AI systems to improve themselves, pursue power, grab resources, and resist being turned off or having their goals changed and that, therefore, we should expect sufficiently advanced, goal-directed AI systems to exhibit these problematic behaviours, has long been recognised by computer scientists. In their popular AI textbook, Stuart Russell and Peter Norvig (2020, 1842), relay that AI pioneer Marvin Minsky "once suggested that an AI programme designed to solve the Riemann Hypothesis might end up taking over all the resources of Earth to build more powerful supercomputers." The classic reference is Omohundro (2008), and Bostrom (2012) discusses similar issues, such as the "instrumental convergence thesis."

80. Other books on the risks posed by AGI include Christian (2021); Russell (2019); and Tegmark (2017).

81. Some of these scenarios are discussed in *Superintelligence*, too (Bostrom 2014b). Some of the most illuminating recent discussions about AI risk have not been formally published but are available online—see, for instance, Ngo (2020); Carlsmith (2021); Drexler (2019), and the work of AI Impacts (https://aiimpacts.org/). For an overview of different ways in which an AGI takeover might happen, see Clarke and Martin (2021).

82. The AI Alignment Forum (https://www.alignmentforum.org/) is a good place to follow cutting-edge discussions on AI alignment. For a recent conceptual overview of the field, see Christiano (2020). Different authors have used different ways of conceptualizing the challenge of creating AI systems that are more capable than humans but lead to desirable outcomes when deployed. Yudkowsky (2001) described the issue as how to create "friendly AI"; Bostrom as the "control problem" (Bostrom 2014b, Chapter 9). (See also Christiano 2016, 2018a; Gabriel 2020; Hubinger 2020.)

83. What about worlds that are controlled by AIs but without significant lock-in? We can, for example, imagine a society of AIs that reflect, reason morally, and remain open to changing their minds. At present I have little to say about such scenarios because I'm

uncertain how to evaluate them. I feel clueless about whether to expect better or worse results from this society than from a world tethered to human values. See also Christiano 2018b.

84. Haldane 1927. More at whatweowethefuture.com/notes.

85. I thank Thomas Moynihan for pointing me to this essay. Haldane made some major, and less forgivable, errors in other areas too. He was a proponent of eugenics, and in 1962 he described Stalin as "a very great man who did a very good job" (R. W. Clark 2013, Chapter 13). Haldane's vision in "The Last Judgment" of how humanity would settle outer space—first Venus, then the Milky Way and beyond—is disturbing as well, arguably an example of flawed value lock-in: individual liberties and regard for happiness, art, and music are described as "aberrations" that nearly caused humanity's extinction; only a large-scale eugenics effort allows some humans to escape to Venus, where "the evolution of the individual has been brought under complete social control" and, because of a new perceptual sense, "every individual at all moments of life, both asleep and awake, [is] under the influence of the voice of the community" (foreshadowing the Borg from *Star Trek*). Other scientists were also poor at predicting space travel. In 1957, Lee de Forest, an American radio pioneer and inventor of the triode vacuum tube, predicted that we would never land on the moon (*Lewiston Morning Tribune* 1957).

86. "For several decades the costs of solar photovoltaics (PV), wind, and batteries have dropped (roughly) exponentially at a rate near 10% per year. The cost of solar PV has decreased by more than three orders of magnitude since its first commercial use in 1958" (Way et al. 2021, 2). The text's Figure 1 exhibits a relatively constant PV cost decline since about 1960.

87. "Most energy-economy models have historically underestimated deployment rates for renewable energy technologies and overestimated their costs" (Way et al. 2021, 1). On photovoltaics (PV) specifically, they present "a histogram of 2,905 projections by integrated assessment models, which are perhaps the most widely used type of global energy-economy models, for the annual rate at which solar PV system investment costs would fall between 2010 and 2020. The mean value of these projected cost reductions was 2.6%, and all were less than 6%. In stark contrast, during this period solar PV costs actually fell by 15% per year. Such models have consistently failed to produce results in line with past trends. . . . In contrast, forecasts based on trend extrapolation consistently performed much better" (3f).

88. Cotra 2020. For a summary, see Karnofsky (2021d). Technically, Cotra considers the training requirements for what she calls a "transformative model," which she defines as a neural network constituting a "single computer program which performs a large enough diversity of intellectual labor at a high enough level of performance that it alone can drive a transition similar to the Industrial Revolution," that transition requiring the economic growth rate to increase by a factor of ten, from 2–3 percent to 20–30 percent per year. While this is conceptually different from my definition of AGI, I believe that for our purposes we can use these concepts roughly interchangeably: On one hand, I believe that AGI would be sufficient to cause an Industrial Revolution–scale growth acceleration, as I discuss later in this chapter. On the other hand, I think that a transformative model would either very quickly lead to the development of AGI or have similar implications as AGI, including for value lock-in.

89. "Today's AI systems are sometimes as big as insect brains, but never quite as big as mouse brains—as of this writing, the largest known language model was the first to come reasonably close—and not yet even 1 percent as big as human brains" (Karnofsky 2021d).

90. The amount of computing operations used in the largest AI training runs doubled every 3.4–3.6 months between 2012 and 2017, increasing by a factor of three hundred thousand over that period (Amodei and Hernandez 2018; Heim 2021). Since then, the trend has slowed: a follow-up analysis of the period 2012–2021 found a doubling time of 6.2

months. Note that, over a decade, this still corresponds to an increase by a factor of more than 670,000. (See also AI Impacts, n.d.-d, n.d.-a; Hernandez and Brown 2020; Moore 1965; Supernor 2018.)

91. "In the coming decade or so, we're likely to see—for the first time—AI models with comparable 'size' to the human brain" (Karnofsky 2021d). On Cotra's "best guess" assumptions, the chance that we'll have enough computing power for AGI by 2100, conditional on what she calls the "Evolution Anchor," is a bit over 50 percent. See Cotra 2020, Part 4, 9.

92. It is worth distinguishing two types of uncertainty involved in Cotra's model (and indeed any model). Cotra discusses several different ways of comparing AI systems to biological systems and calls these different ways of comparison "biological anchors." The first type of uncertainty is the one acknowledged in the main text: conditional on each biological anchor, we might over- or underestimate the amount of computing power required to train AGI. Uncertainties of this are represented within the model as probability distributions, and their effects can be combined into a single bottom-line probability distribution that allows for statements like "a 50 percent chance of AGI by 2050." But, crucially, any such statement only takes into account this type of uncertainty. The second type of uncertainty is uncertainty about parameters that within the model are represented as single numbers rather than probability distributions. Important examples of such parameters are the weights assigned to each biological anchor—essentially the assumed probability that each particular anchor correctly predicts the computing power requirements for training AGI. For instance, the result of "a 50 percent chance of AGI by 2050" is based on assigning a weight of 10 percent to the Evolution Anchor. If you think the Evolution Anchor is less likely (or more likely) to be "correct," then your version of Cotra's model would predict a chance of AGI by 2050 that's different from 50 percent. To make our uncertainty of the second type visible, we need to compare how the model output changes for different assumptions about its parameters. The probabilities stated in the main text express the uncertainty of the first type conditional on Cotra's best-guess assumptions about parameter values ("I am tentatively adopting ~2050 as my median forecast for TAI," Part 4, 15; and "~12%–17%" for 2036, Part 4, 16). On Cotra's (2020) "conservative" assumptions, the results instead are 50 percent by 2090 (Part 4, 15) and 2–4 percent by 2036 (Part 4, 16); on her "aggressive" assumptions, the results are 50 percent by 2040 (Part 4, 15) and 35–45 percent by 2036 (Part 4, 16). The difference between conservative, best-guess, and aggressive assumptions is due to uncertainty of the second type. You can explore how the results of the model differ by putting your own assumptions in a Colab notebook and spreadsheet which are available online (Cotra, n.d.).

93. Wiblin and Harris 2021, January 19. The quoted parts appear at time stamps 1:33:38 and 1:35:38 of the podcast, respectively.

94. Grace et al. 2018. In 2019, the Centre for the Governance of AI conducted a follow-up survey containing many of the same questions; the results, publication of which is forthcoming, broadly confirm the findings I described in the text (B. Zhang et al. 2022). For an (incomplete) overview of other AI timeline surveys, see AI Impacts (n.d.-b), and for an overview including predictions by individuals, see Muehlhauser (2016a).

95. More precisely, the "survey population was all researchers [*n* = 1,634] who published at the 2015 NIPS and ICML conferences" (Grace et al. 2018, 730). Of these, *n* = 352 researchers responded, yielding a response rate of 21 percent.

96. Grace et al. 2018, 730, 736.

97. Grace et al. 2018, 731.

98. Grace et al. 2018, 732, Figure 2.

99. "The peak of AI hype seems to have been from 1956–1973. Still, the hype implied by some of the best-known AI predictions from this period is commonly exaggerated" (Muehlhauser 2016b; an extended discussion of this assessment is in the same work). For a history of AI as a research field, see, e.g., Nilsson (2009).

100. "pr(AGI by 2036) ranges from 1% to 18%, with my central estimate around 8%" (Davidson 2021a).

101. Pew Research, n.d.; Pew Research 2014.

102. Buddhism started fading in Afghanistan with the Muslim conquest in the seventh century, but Islam only took over the main cities of Afghanistan in AD 900, and some remote regions held on to their native religion until the nineteenth century. Zoroastrianism, Hinduism, and paganism also had many adherents throughout Afghan history (Azad 2019; Green 2016, Introduction; Runion 2007).

103. Benjamin 2021; *Encyclopedia Britannica* 2018a; H. P. Ray 2021; *Encyclopedia Britannica* 2020g; Green 2016.

104. CIA 2021.

105. "The Comintern functioned chiefly as an organ of Soviet control over the international communist movement" (*Encyclopedia Britannica* 2017). "The Seventh Congress of the Comintern [was] incidentally the last to take place" (Rees 2013).

106. Our World in Data, n.d.-a.

107. Yglesias 2020.

108. If we could capture all the sun's solar energy that hits Earth, we would be able to capture 1.3×10^{17} W. If we put a Dyson sphere around our sun, we could capture 4×10^{26} W, which is three billion times as much.

The Milky Way has roughly a hundred billion stars (Murphy 2021, Section 1.2). Tapping into this abundance of energy would quickly solve all of the problems that derive from energy scarcity, such as food production, water purification, and conflicts over oil. We could also get additional resources by mining asteroids and neighbouring planets (Ord 2020, 227f).

109. Stark 1996.

110. Stark 1996, 4–13.

111. Stark 1996, 7.

112. Pew Research 2015. For the definition of "religiously unaffiliated," see Appendix C of that text.

113. Pew Research 2015.

114. World Bank 2021f; Roser et al. 2019.

115. World Bank 2021c; Gramlich 2019.

116. Gramlich 2019. The claims in this paragraph are based on the UN's population projections. As I explain in a note in Chapter 7, I'm more persuaded by the forecast from Vollset et al. (2020), in which the effects I mentioned would be even bigger. More online.

117. Wood et al. 2020.

118. "Within 50 years following contact with Columbus and his crew, the native Taino population of the island of Hispaniola, which had an estimated population between 60,000 and 8 million, was virtually extinct (Cook, 1993)" (Nunn and Qian 2010, 165).

119. Although most countries are moving towards Western values, they are moving towards Western values at different speeds, so in some cases values are diverging, not converging. However, if trends continue, at some point most countries will converge on Western values because there must be a limit on how "Western" a country can become (Kaasa and Minkov 2020).

120. This argument has also been made by Hanson (2020).

121. BioNTech 2021; Moderna 2021.

122. Cochrane 2020. Some countries did allow vaccines to be bought on the free market after they were tested (Menon 2021).

123. While Japan invaded China in 1937, World War II is generally considered to have started with Nazi Germany's attack on Poland on September 1, 1939.

Regarding Hitler's international prestige: On Nazi sympathies in the United States, see, for instance, Hart (2018, 27), who contends that "given how far Nazism managed to spread on its own in the United States, it was fortunate that the Germans were not more adept at pressing their advantages." One of the most infamous Hitler sympathizers in Britain was *Daily Mail* cofounder Harold Sidney Harmsworth, 1st Viscount Rothermere, who met and corresponded with Hitler multiple times in the 1930s (Kershaw 2005). More at whatweowethefuture.com/notes.

124. The following argument is also made in an excellent article by Evan Williams (2015).

125. You might be balking at the idea that there is such a thing as a "morally best" society. I'm not, here, wedding myself to the idea that there is a single objective moral truth, though I think that idea has more going for it than some would believe. But I am claiming that moral views can be better or worse: that proslavery moral views are worse than antislavery moral views; that it's incorrect to think that torturing children is admirable. One way of understanding this, without committing oneself to the spooky metaphysics of objective moral truths, is to think of the morally correct view as the moral view that you would come to endorse if you had perfect information and unlimited time to reflect, could experience a diversity of lives, and were exposed to all the relevant arguments.

126. A common myth is that Shenzhen grew from a small fishing village to a huge city over the course of a few decades, but this isn't true. In 1979, Shenzhen was a market town with some industry and a population of 310,000 (Du 2020, Chapter 1). Special economic zones have been tried in other places, but in spite of some successes like Shenzhen, on average, they have not grown faster than their host country (Bernard and Schukraft 2021).

127. In 1980, per capita income was $122, and in 2019, it was $29,498 (Charter Cities Institute 2019; *China Daily* 2020; Yuan et al. 2010, 56).

128. Roser and Ortiz-Ospina 2017; Yuan et al. 2010.

129. Esipova et al. 2018.

130. Toby Ord (2020) gives another example of this paradox in *The Precipice*. He suggests we should perhaps lock in a commitment to avoiding our own extinction or other terrible outcomes for humanity but, for now at least, should try not to lock in more than that.

131. Forst 2017. See also Belfield (forthcoming).

132. For related worries about what would happen if the future was shaped by the unchecked forces of biological and cultural evolution, see Bostrom (2004).

Chapter 5: Extinction

1. Alvarez et al. 1980; Wignall 2019a, 90–91.

2. Chapman 1998.

3. NASA 2021; Crawford 1997. The total yield of the world's nuclear arsenal in 2019 was around 2.4 billion tonnes (estimated by van der Merwe [2018] using data from Kristensen et al. [2018]; Kristensen and Korda [2018, 2019a, 2019b, 2019c, 2019d]; Kristensen and Norris [2011, 2017]).

4. NASA 2019.

5. Asay et al. 2017, 338.

6. S. Miller 2014.

7. *Science* 1998. The Shoemaker-Levy comet was jointly discovered by David Levy, Carolyn Shoemaker, and Gene Shoemaker (Carolyn's husband).

8. Chapman 1998. In the DVD commentary to *Armageddon* (Bay 1998), Ben Affleck said that he asked director Michael Bay "why it was easier to train oil drillers to become astronauts than it was to train astronauts to become oil drillers": "He told me to shut the fuck up, so that was the end of that talk" (servomoore 2016).

9. A. Harris 2008.

10. Clarke 1998.

11. A. Harris and Chodas 2021, 8.

12. Alan Harris, personal communication, October 4, 2021.

13. Ord 2020, 71; Alan Harris, personal communication, October 4, 2021.

14. Newberry 2021.

15. This is an estimate by the *Economist* of the excess deaths from COVID-19 up until November 22, 2021 (*Economist* 2021c). While 17 million excess deaths are the stated best guess, there is considerable uncertainty: the estimate indicates that with 95 percent probability the true number falls between 10.8 million and 20.1 million.

Excess deaths measure the difference between how many people died during the COVID-19 pandemic compared to an estimate of how many would have died if COVID-19 had not happened. This accounts for various issues of under- and overreporting of deaths attributable to COVID-19. More at whatweowethefuture.com/notes.

16. *Economist* 2021b.

17. Wetterstrand 2021; BC 2018, Figures 6 and 7; Boeke et al. 2016, Figure S1 A, page 2 of the Supplementary Materials. On Moore's law in terms of cost, see Flamm (2018).

18. Wetterstrand 2021.

19. Ord 2020, 137.

20. Nevertheless, many governments have successfully concealed their nuclear weapons programmes, though this is somewhat harder to do if countries also pursue civilian nuclear power (Miller 2017).

21. Anderson 2002, 49.

22. Anderson 2002, 10.

23. Anderson 2002, 5, 8.

24. The company working to develop the vaccines for foot-and-mouth disease was called Merial Animal Health, but we cannot completely rule out the possibility that the leak could have come from the Pirbright Institute of Animal Health. Merial was based at the Pirbright Institute, which was also researching foot-and-mouth disease. Major government reports on the outbreak concluded that the outbreak likely came from Merial because Merial produced far more of the foot-and-mouth virus (Spratt 2007, 5, 10).

25. Anderson 2002, 11.

26. Spratt 2007, 9.

27. Anderson 2008, 8, 11.

28. Anderson 2008, 107.

29. Manheim and Lewis 2021, Table 1; Okinaka et al. 2008, 655; Tucker 1999, 2.

30. Alibek and Handelman 2000, 74.

31. Zelicoff 2008, 106–108.

32. Bellomo and Zelicoff 2005, 101–111.

33. It is disputed whether the woman was asymptomatic or not. The woman in question, Bayan Bisenova, said she was, but the Soviets claimed she had started experiencing symptoms (Zelicoff 2003, 105).

34. Zelicoff 2003, 100.

35. Furmanski 2014.

36. Hansard 1974.

37. Shooter 1980.

38. National Research Council 2011, Table 2.6.

39. National Research Council 2011, 34, Table 2.6.

40. During the Cold War, the Soviets devised a similar system for nuclear weapons, known informally as the "Dead Hand," that would allow them to launch a nuclear counterstrike even if a US first strike obliterated their command centres (Ellsberg 2017, Chapter 19; Hoffman 2013).

41. Carus 2017b, 144.

42. Carus 2017b, 139, 143.

43. Carus 2017b, 148.

44. Carus 2017b, 146; Ouagrham-Gormley 2014, 96.

45. Carus 2017b, 147.

46. Carus 2017b, 129–153; Meselson et al. 1994; Ouagrham-Gormley 2014; P. Wright 2001.

47. Lipsitch and Inglesby (2014) estimate that there is one accidental infection per 100 full-time employees. However, they use a small sample, and once we use a larger sample (National Research Council 2011, 34, Table 2.6), a figure of one infection per 250 employees becomes more plausible. Professor Lipsitch agreed in an email communication on October 3, 2021, that the larger sample should be used.

48. Shulman 2020.

49. See, for example, Alibek and Handelman 2000, 198. However, Alibek is often cited as an unreliable witness (Leitenberg et al. 2012, 7).

50. Manheim and Lewis 2021, 11.

51. Michaelis et al. 2009, Table 1; Nakajima et al. 1978; Rozo and Gronvall 2015; Scholtissek et al. 1978; Wertheim 2010; Zimmer and Burke 2009. Michaelis et al. (2009) do not provide a source for their estimate of the number of people killed in the Russian flu pandemic, so I am unsure how reliable it is, and I have been unable to find other official estimates.

52. S. H. Harris 2002, 18f.

53. L. Wright 2002.

54. Leitenberg 2005, 28–42.

55. This can be inferred from estimates for a series of three questions on Metaculus: (1) "By 2100 will the human population decrease by at least 10% during any period of 5 years or less ["global catastrophe"]?"; (2) "If a global catastrophe happens before 2100, will it be principally due to . . . bioengineered organisms?"; and (3) "Given [the former], will the global population decline more than 95% relative to the pre-catastrophe population?" (Tamay 2019). As of November 18, 2021, the combined forecasts for these events put the risk of a pandemic killing at least 95 percent of people at 0.6 percent. The estimates will likely change in the future.

56. Ord 2020, 71.

57. The real risk that a plane will crash is less than one in a million (UK Civil Aviation Authority 2013).

58. NASA 2021.

59. In addition to asteroids, comets, and engineered pathogens, there are many other natural and anthropogenic extinction risks. These include supervolcanic eruptions, gamma ray bursts, nuclear war, and climate change. The extinction risk these threats pose is discussed at length by Ord (2020). I discuss the risks from nuclear war and climate change in Chapter 6.

60. The term "Long Peace" was first coined in 1986 by John Lewis Gaddis in an article that noted a systemic absence of war, not just an absence of great-power wars (Gaddis 1986). More recently, in The Better Angels of Our Nature, psychologist Steven Pinker argued that there has been a longterm decline in war, especially since World War II, as part of a general civilisational decline in violence of all kinds (Pinker 2011). Political scientists like John Mueller (2009) and Azar Gat (2013, 149) have made similar points.

61. One database, compiled by the Future of Life Institute, counts at least twenty-five close calls during the Cold War (Future of Life Institute, n.d.).

62. Pinker 2011, 208.

63. International relations scholar Bear Braumoeller has calculated that, if the annual chance of a "systemic" war breaking out is 2 percent, then there's a roughly 25 percent chance that a given seventy-year period is peaceful (Braumoeller 2019, 26–29). Statisticians Pasquale Cirillo and Nassim Taleb have shown similarly that long periods of peace are statistically compatible with a constant risk of war (Cirillo and Taleb 2016ab).

64. World Bank 2021h.

65. Power transition theory was pioneered by the political scientist A. F. K. Organski in 1958 and has been an active field of research since. In his summary of the evidence for various causes of war, political scientist Greg Cashman (2013, 485) writes, "Serious great-power crises have in the past been most likely to occur during periods of transition in the international system (or in regional subsystems) where there are significant shifts in the balance of capabilities, especially between the dominant power in the system and its major rival(s)." For a recent overview of the theory, see Tammen et al. (2017).

66. See Cashman 2013, 416–418. Cashman finds that estimates of the base rate of conflict during power transitions vary depending on the data and methods used but are as high as 50 percent. However, it's worth noting that there is some evidence suggesting that future power transitions may pose a lower risk of war, not an elevated one, and some researchers believe that it is equality of capabilities, not the transition process that leads to equality, that raises the risk of war.

67. See whatweowethefuture.com/notes.

68. "Historically, large, powerful states have been more likely to be involved in war than small, less powerful states" (Cashman 2013, 479).

69. Bulletin of the Atomic Scientists 2021.

70. See Our World in Data 2019g, 2019f. Those sources are based on UN 2019b.

71. India reported twenty fatalities as a result of the conflict. China did not reveal how many losses its forces suffered, but one report, citing US intelligence estimates, claimed thirty-five Chinese soldiers died (*US News* 2020). Most of the casualties occurred when soldiers, fighting at night in treacherous conditions, fell to their deaths from the high mountain pass (*Guardian* 2020).

72. Gokhale 2021.

73. Cashman (2013, 478–479) writes that there is general agreement among social scientists that interstate wars "almost always" occur between neighbouring countries. A territorial dispute is the issue most likely to spark a war. The other patterns are the following:

- "Large power disparities between states seem to promote peace rather than war."
- A "disproportionately large percentage of wars involve . . . strategic rivals"—that is, states with "an extended mutual history of hostile interactions that probably include participation in serial crises and/or militarized disputes with each other, and perhaps even a history of previous wars."
- Large, powerful states are more likely to fight than small, less powerful states; "mature democracies" are "highly unlikely to ever fight each other."
- Most wars are "preceded by militarized disputes or crises that involve escalatory behavior preceding the outbreak of war that looks like a conflict spiral."

74. Per data available in World Bank 2021n. It's worth noting, though, that the strength of the effect of economic interdependence on the likelihood of war is far from clear and is disputed by some scholars (Levy and Thompson 2010, 70–77).

75. Waltz 1990.

76. See, e.g., Tannenwald 1999.

77. Jgalt 2019.

78. Historian Ian Morris (2013, 175) has attempted to quantify humanity's war-making capacity, defined as "the number of fighters they can field, modified by the range and force of their weapons, the mass and speed with which they can deploy them, their defensive power, and their logistical capabilities." He estimates that this measure increased by a factor of between fifty and one hundred over the course of the twentieth century. It's very likely that advances in areas like automation, biotechnology, and military science will drive further increases in the future. Bear Braumoeller, in Chapter 5 of *Only the Dead*, analyses longterm trends in the deadliness of international conflict. At the end of the chapter he

writes, "When I sat down to write this conclusion I briefly considered typing, 'We're all going to die,' and leaving it at that. . . . If the parameters that govern the mechanism by which wars escalate hasn't changed—and there's no evidence to indicate that they have—it's not at all unlikely that another war that would surpass the two World Wars in lethality will happen in your lifetime" (Braumoeller 2019, 130).

79. Rose 2006, 50. Estimates of when chimps and humans split differ, ranging from 5.7 million years ago (Reis et al. 2018, Table 1, Strategy B, Minimum) to 12 million years ago (Moorjani et al. 2016); more at whatweowethefuture.com/notes.

80. Schlaufman et al. 2018. Krauss and Chaboyer (2003) give an estimate of 13.4 billion years.

81. Bostrom 2002.

82. Los Alamos National Laboratory 2017.

83. Bostrom 2002.

84. Sandberg et al. 2018. The model used by Sandberg et al. (2018) has been criticised by James Fodor (2020). More at whatweowethefuture.com/notes.

85. The earth became cool enough for life around 4 billion years ago, with uncertainty on the order of hundreds of millions of years (Knoll and Nowak 2017, Figure 1). The earth will become uninhabitable in around 0.8 to 2 billion years (Lenton and von Bloh 2001; O'Malley-James et al. 2013; Ord 2020, 221–222; von Bloh 2008; Wolf and Toon 2014).

86. Hanson et al. 2021.

87. Hanson (1998) says his model might be compatible with anywhere between one and seven hard steps.

Chapter 6: Collapse

1. Scheidel 2021, 102, Figure 7, and 103, Table 2.2. More at whatweowethefuture.com/notes.

2. Ionescu et al. 2015, 244.

3. Jackson et al. 2013, 2017.

4. National Geographic Society 2018; *Encyclopedia Britannica* 2011.

5. The Roman Empire controlled at least four million square kilometres and probably over five million, depending on how much desert is included (Scheidel 2019, 34). The land area of the European Union is just below four million square kilometres (World Bank 2021i).

6. Temin 2017, Chapter 8; G. K. Young 2001. There is evidence that the Roman Empire traded with the Korean Empire (UNESCO, n.d.).

7. Petronius satirised the newly rich in the character of Trimalchio in the *Satyricon*, written during Nero's reign in the first century AD. Scheidel and Friesen (2009, 84–85) estimate that around 10 percent of the population would have enjoyed "middling" incomes, "defined by a real income of between 2.4 and 10 times 'bare bones' subsistence or 1 to 4 times 'respectable' consumption levels."

8. Ward-Perkins 2005, 94f.

9. Morris 2013, 147–148, Table 4.1, and 155–156, Table 4.2. This estimate comes with the qualification that ancient demography is a very uncertain affair.

10. Scheidel 2019, 81f.

11. Jerome, *In Ezekiel*, I *Praef.* and III *Praef.* (Migne, *Patrologia Latina* XXV, coll. 15–16, 75D): "in una Urbe totus orbis interiit."; quoted in Ward-Perkins 2005, 28.

12. See whatweowethefuture.com/notes.

13. Morris 2013, 151. Rome's peak population was about one million from AD 1 to 200 and, according to Morris, did not reach that peak again until the twentieth century (Morris 2013, 147–148, Table 4.1). The city of Rome didn't have a population larger than one million until the 1930s (Ufficio Di Statistica E Censimento 1960).

14. Morris 2013, Table 4.1.

15. Cited in Scheidel 2019, 128.

16. Scheidel 2019, 129.

17. Scheidel 2019, Chapter 5.

18. The exact figure is 336 years (Stanaway et al. 2018).

19. Ward-Perkins 2005, 164.

20. Ward-Perkins 2005, 108.

21. Walter Scheidel makes this argument at length in *Escape from Rome*, where he discusses the many other proponents of this theory (Scheidel 2019, 538n19).

22. National Geographic Society 2021; *Encyclopedia Britannica* 1998, 2021f, 2020e, 2020c, 2019c.

23. This depends on the data source used. One piece of World Bank data suggests that world GDP has fallen relative to the previous year six times since 1960 and has always passed the previous peak within two years (World Bank 2021d). However, other sources suggest that GDP declined only four times in the last hundred years: 1930–1932, the Great Depression; 1945–1946, World War II; 2009, the Great Recession; and 2020, the start of the COVID-19 pandemic (IEA 2020b, using the 2020 Maddison database [Bolt and van Zanden 2020] and Geiger's [2018] interpolations from the 2014 Maddison database [Bolt and van Zanden 2014]).

24. Roser 2020a. The lead author of a recent study estimating the death toll of the Spanish flu told us that he doesn't believe there was a population decline in that year (Spreeuwenberg et al. 2018, personal correspondence, August 18, 2021).

25. Human Security Project 2013, 36f; Roser et al. 2019.

26. G. Parker 2008; Zhang et al. 2011.

27. Zhang et al. 2011.

28. Zhang et al. 2011, 297; G. Parker 2008, 1059.

29. Ord 2020, 349f.

30. Ord 2020, 124.

31. Ord 2020, 350. Some economic historians even argue that the Black Death sped up subsequent economic growth. In the century that followed, European wages more than doubled; one argument is that, because so many people died, there was a lot more land per person. This increased the value of labour relative to land, giving greater incentives for investment in capital accumulation and innovation (Clark 2016).

32. The bomb dropped on Hiroshima was fifteen thousand tonnes of TNT equivalent (Malik 1985). The largest conventional bomb dropped during World War II was the Grand Slam, which was around ten tonnes of TNT equivalent (*Encyclopedia Britannica* 2021d).

33. *Encyclopedia Britannica* 2021d; Lifton and Strozier 2020; US Strategic Bombing Survey 1946.

34. US Department of Energy, n.d.

35. Wellerstein 2020.

36. Hiroshima Peace Memorial Museum, n.d.

37. McCurry 2016.

38. Chugoku Shimbun 2014. For differing accounts, see whatweowethefuture.com/notes.

39. US Department of Energy, n.d.; Kuwajima 2021; Wada 2015.

40. Hiroshima Convention and Visitors Bureau, n.d.

41. Population estimates of Hiroshima prior to the bombing differ, with some putting the number at 255,000 and others putting it at 343,000 (*Encyclopedia Britannica* 2021d; French et al. 2018). The population had reached 357,000 by 1955 (UN 1963, 341).

42. Center for Spatial Information Science 2015.

43. D. R. Davis and Weinstein 2008, 38.

44. D. R. Davis and Weinstein 2008.
45. Miguel and Roland 2011.
46. Dartnell 2015a, 47f.
47. Dartnell 2015a, 193.
48. Cochran and Norris 2021.
49. Wellerstein 2021.
50. Roser and Nagdy 2013.
51. Ord 2020, 26.
52. Ord 2020, 96f.
53. Roser and Nagdy 2013.
54. See whatweowethefuture.com/notes.
55. Some studies suggest that a Russian attack on the United States would kill tens to hundreds of millions of people, depending on the targeting strategy. The global death toll of an all-out war would be higher, but these numbers need to be adjusted for higher population and smaller arsenals (Helfand et al. 2002; Ord 2020, 334n24). Luisa Rodriguez (2019) estimates that with current arsenals, an all-out Russia-NATO nuclear war would lead to fifty-one million fatalities.
56. Coupe et al. 2019, Figure 7; Robock et al. 2007, Figure 2.
57. Coupe et al. 2019, Figures 10, 12.
58. Robock 2010. Note that these nuclear winter models are controversial, and some models suggest that the cooling would be considerably smaller. The possibility of nuclear winter has been controversial since it was first proposed in the 1980s (see, e.g., Maddox 1984; Penner 1986). Reisner et al. (2018) have criticised estimates of nuclear winter using modern climate models.
59. IFLA 2021. More at whatweowethefuture.com/notes.
60. Roser 2013c; Rapsomanikis 2015, 9. About two-thirds of the developing world's three billion rural people live in about 475 million small farm households, working on land plots smaller than two hectares.
61. Robock et al. 2007.
62. Coupe et al. 2019, Figure 9.
63. Shead 2020.
64. Roser and Ritchie 2013; Ritchie and Roser 2020b; US Energy Information Administration 2021a. More at whatweowethefuture.com/notes.
65. See Belfield (forthcoming) and whatweowethefuture.com/notes.
66. This illustrates that low population itself does not imply civilisational collapse, but, as Matthew van der Merwe has pointed out to me, the comparison is not perfectly analogous because there might be an important difference between starting out with a low population and having a low population because of a massive catastrophe. The last time I weighed twenty kilograms I was six years old, and being at such a weight was no risk to my health. But if my weight dropped to twenty kilograms now, I would surely die.
67. Doebley et al. 1990, Figure 2.
68. Renner et al. 2021; National Science Foundation 2020.
69. Dartnell 2015a, 52f.
70. Allard 2019.
71. Barclay 2007; Engelen et al. 2004.
72. Barclay 2007; Gupta et al. 2019; Perez et al. 2009; Whitford et al. 2013.
73. Balter 2007.
74. Balter 2007.
75. Richerson et al. 2001.
76. It is true that we do not really know how long it would have taken different civilisations to industrialise had they been isolated from European influence and colonialism. In AD 1500,

even though they had had agriculture for thousands of years, the Americas were not close to having industrial technology. We do not know when, or even if, they would have industrialised had they not been colonised by Europeans. Perhaps Native American societies were in a different equilibrium and did not pursue industrialisation, or perhaps industrialisation is very difficult to achieve. Still, given that knowledge of industrial processes would very likely still be available in the postcollapse world, on balance, it seems like there would be fewer barriers to industrialisation for a postcollapse society that was seeking to reindustrialise.

77. Many concrete buildings from ancient Rome have survived, but modern reinforced concrete is not actually very durable and will start to degrade after only twenty years (Alexander and Beushausen 2019; Daigo et al. 2010).

78. Daigo et al. 2010.

79. I'm here echoing sentiment from Bill McKibben (2021).

80. IEA 2020a, 195.

81. Hausfather 2021b; US Energy Information Administration 2021b.

82. Hausfather 2020.

83. Kavlak et al. 2018; Sivaram 2018, Chapter 2; Roser 2020b; Ritchie 2021.

84. Ritchie and Roser 2020b.

85. McKerracher 2021, Figure 2.

86. Mohr et al. 2015; Welsby et al. 2021, SI section 2.

87. See whatweowethefuture.com/notes.

88. Most of the climate-impacts literature focuses on the impact of an extreme emissions scenario known as "RCP8.5," in which there would be between four and five degrees of warming by the end of the century (Hausfather and Peters 2020).

89. Buzan and Huber 2020, Figure 10; Prudhomme et al. 2014.

90. Sloat et al. 2020; Zabel et al. 2014. The IPCC finds that five degrees of local warming in temperate regions has close to zero effect on yields (IPCC 2014b, 498). Moreover, yields for the major food crops have increased by a factor of two to three over the last sixty years (H. Ritchie and Roser 2021).

91. Buzan and Huber 2020.

92. For example, Ramirez et al. (2014) find that "on the most alarmist assumptions possible," their model nearly runs away at 3,300 parts per million, a level of carbon dioxide concentration that is probably out of reach from recoverable fossil fuels (see also Goldblatt and Watson 2012; Wolf and Toon 2014).

93. Hansen et al. 2013, 17. Popp et al. (2016) found that if carbon dioxide concentrations reached 1,520 parts per million, a simulated planet would transition to a moist greenhouse state. If we burned all of the fossil fuels, then carbon dioxide concentrations would reach 1,600 parts per million (Lord et al. 2016, Figure 2). However, the simulated planet's initial climate was six degrees warmer than today's Earth. This means that Earth would require a carbon dioxide concentration significantly higher than on the simulated planet to transition to a moist greenhouse. More at whatweowethefuture.com/notes.

94. The model found that the warming would happen over the course of a month, but in reality the transition would take longer (Schneider, personal communication, August 20, 2021; Schneider et al. 2019). More at whatweowethefuture.com/notes.

95. Lord et al. 2016, Figure 2. More at whatweowethefuture.com/notes.

96. Hausfather 2019; Voosen 2019.

97. Foster et al. 2017, Figure 4.

98. Lethal limits for the major food crops are between forty and fifty degrees Celsius (King et al. 2015). Although some places in the tropics would pass these limits for part of the year with fifteen degrees of warming, North America, Europe, and China would not.

99. Climate change could also be a stressor for other catastrophic risks, such as the risk of war. The effect of climate change on conflict is very controversial; there is some evidence

linking climatic changes to increased levels of civil conflict in Africa, although most conflict researchers believe that it is a small driver relative to other factors, such as state capacity and economic growth. For contrasting takes on the climate and conflict connection, see Buhaug et al. (2014) and Hsiang et al. (2013). For a survey of leading climate and conflict researchers, see Mach et al. (2019).

100. Lord et al. 2016; Talento and Ganopolski 2021. More at whatweowethefuture.com/notes.

101. The loss of knowledge after a drop in population size is known as the Tasmania effect. More at whatweowethefuture.com/notes.

102. There are several important exceptions to this. For example, Argentina and Brazil both initially relied mainly on hydropower, oil, and gas rather than coal, while the Philippines relied mainly only on oil and then shifted to other energy sources (Ritchie and Roser 2020b).

103. Dartnell 2015b. See also Belfield (forthcoming).

104. Davis et al. 2018.

105. Dartnell 2015b.

106. J. Ritchie and Dowlatabadi 2017; Rogner et al. 2012, Section 7.4.

107. Rogner et al. 2012, Table 7.18.

108. Rogner et al. 2012, Table 7.18.

109. Banerjee 2017; BNSF Railway 2018, 14.

110. Between 1800 and 1850, the world used forty-four exajoules of energy (Ritchie and Roser 2020b). The nine hundred million tonnes of carbon in coal at North Antelope Rochelle is equivalent to around twenty-four exajoules.

111. As of 2010, there were 7,800 exajoules of energy remaining in surface coal reserves (Rogner et al. 2012, Table 7.18). Between 1800 and 1980, we used around 7,400 exajoules from fossil fuels (Ritchie and Roser 2020b).

112. US Energy Information Administration 2021a. For all countries except the United States, the most recent data on surface reserves is from Rogner et al. (2012, Table 7.18). For surface coal production data, see Elagina (2021); Geoscience Australia (2016); Huang et al. (2017); Mukherjee and Pahari (2019); US Energy Information Administration (2021a).

113. L. Roberts and Shearer 2021. We should be uncertain about the future demand for coal. Thus far, part of the decline in coal demand has been driven by the declining cost of natural gas from fracking. However, over the last century, the costs of both coal and gas have fluctuated within a fairly narrow range. Empirically informed cost projections suggest that the costs of coal and gas will not change much in the future, so it is unclear whether switching from coal to gas will continue, especially as global demand for gas increases (Way et al. 2021, Figure 3).

114. See whatweowethefuture.com/notes.

115. The precise share of hard-to-replace emissions is 27 percent (Davis et al. 2018, Figure 2).

116. Lynas et al., n.d.; Way et al. 2021. More at whatweowethefuture.com/notes.

117. Bandolier 2008.

Chapter 7: Stagnation

1. Baghdad was the capital of the Abbasid Caliphate, which is widely seen as marking the beginning of the Islamic Golden Age (Chaney 2016; *Encyclopedia Britannica* 2020b).

2. Al-Amri et al. 2016, 9; Zhang and Yang 2020, 49; Long et al. 2017; online.

3. Dral-Khalili 2014, Chapters 7 and 8.

4. Dral-Khalili 2014; Hasse 2021; Lyons 2010; Tbakhi and Amr 2007.

5. Scholars disagree about when and to what extent the slowdown in scientific progress in the Islamic world occurred. Some contemporary scholars take the revisionist stance that

progress did not slow down much or that it slowed down later than the twelfth century. More at whatweowethefuture.com/notes.

6. Chaney 2016; Kuru 2019, Part II.

7. Goldstone 2002.

8. Morris (2004) argues that there was substantial growth in per capita incomes in this period, though his estimates seem much too high (pseudoerasmus 2015a, 2015b).

9. For a similar perspective on sustainability, see Bostrom (2014c).

10. Crafts and Mills 2017; raw TFP data from Fernald 2014. Productivity growth briefly sped up again in the late 1990s as information technology boomed. But this turned out to be a temporary upturn, and since then, productivity growth has continued its decline. For the question of whether the apparent decline is just a mismeasurement of recent progress, see whatweowethefuture.com/notes.

11. All of the following is from Gordon (2016) unless otherwise noted.

12. Gordon 2016, 57.

13. O'Neill 2021a; Our World in Data 2019c.

14. Cowen 2018.

15. The sorts of changes that are advocated by those in favour of furthering growth, such as improving the efficiency of scientific institutions, would be very unlikely to change the growth rate permanently (that is, for the full thousand-year period). Our best models of economic growth suggest such permanent "growth effects" are very unlikely; rather, interventions would have a "level effect." That is why I give the example of changing the growth rate from 1.5 percent to 2 percent for a hundred years (which would already be enormously difficult). For more on growth vs. level effects in semiendogenous growth models, our best growth models, see Jones (2005).

16. For work in economic growth theory that explicitly considers timescales of several centuries or more, see, for example, Acemoglu et al. (2005); Galor and Weil (2000); Jones (2001); and Kremer (1993). More broadly, the two types of models that can at least hope to be applicable to such long timescales are known in the literature as endogenous or semiendogenous growth models, respectively. For pathbreaking and Nobel Prize–winning work in this tradition, see Romer (1990); for a recent review, see Jones (2021).

17. For an overview, see Appendix B of Davidson (2021b). In much of the literature, the possibilities of faster-than-exponential and near-zero growth are set aside because they don't fit Kaldor's (1957) "stylised facts" that describe observed growth in the industrial era. For recent exceptions, see Nordhaus (2021) and Aghion et al. (2019) regarding faster-than-exponential growth, and C. Jones (2020) regarding near-zero growth.

18. Technological progress is a necessary condition for sustained economic growth in the models by Solow (1956) and Swan (1956), which are foundational for all of modern growth theory. This is widely acknowledged as a key insight. For example, Jones's popular textbook notes that Solow "emphasized the importance of technological progress as the ultimate driving force behind sustained economic growth" (Jones 1998, 2).

Note, however, that academic economists in the context of growth theory tend to operate with a very broad notion of "technology." For instance, Acemoglu (2008) offers the following words of caution: "Economists normally use the shorthand expression 'technology' to capture factors other than physical and human capital that affect economic growth and performance. It is therefore important to remember that variations in technology across countries include not only differences in production techniques and in the quality of machines used in production but also disparities in productive efficiency ([such as] from the organization of markets and from market failures)" (Acemoglu 2009, 19).

19. The best population projection I'm aware of is one that researchers from the Institute for Health Metrics and Evaluation at the University of Washington produced for the Global Burden of Disease study and published in *The Lancet* (Vollset et al. 2020). They predicted that,

provided that female educational attainment and access to contraceptives continue to increase, world population will very likely "peak just after mid-century and substantially decline by 2100" (1286; see also Figure 5, 1296, which indicates that the predicted decline is approximately exponential). By contrast, the UN's (2019b) widely cited population forecast predicts that population growth will slow down but not stop before 2100; however, Vollset et al. (2020, 1286) argue persuasively that this is based on underestimating the long-run decline in fertility.

20. This is implied by both endogenous and semiendogenous growth models (Jones 2021, 27). For a detailed analysis of a negative-population-growth scenario, see Jones (2020).

21. See whatweowethefuture.com/notes.

22. ATLAS Collaboration 2019; CERN 2017; Cho 2012.

23. Bloom et al. 2020.

24. Based on the data for the aggregate economy, Bloom et al. (2020, Table 7, 1134) estimate a β of roughly 3. This parameter, in a semiendogenous growth model, means that, in equilibrium, a 3 percent increase in research effort yields a 1 percent increase in technological advancement (Bloom et al. 2020, 1135). More detail on why, in the main text, I chose numbers corresponding to a β of 2 is available at whatweowethefuture.com/notes.

25. The example is purely illustrative and intended to gesture very crudely at the kind of innovation that may have been involved in the first doubling of the technology level. For a discussion of which "unit of ideas" is being assumed by the kind of model I rely on here, see Bloom et al. (2020, 1108).

26. Bloom et al. (2021, 1105) find that US research productivity has decreased by a factor of forty-one since the 1930s. The decrease by a factor of five hundred since 1800 is based on a back-of-the-envelope calculation—details at whatweowethefuture.com/notes.

27. Bloom et al. 2020, Figure 1, 1111.

28. The basic observation that a large fraction of all scientists are alive today goes back to at least the "father of scientometrics," Derek de Solla Price (1975, 176), who estimated "some 80 to 90 per cent of all scientists that have ever been, are alive now." See whatweowethefuture.com/notes for why I go for a slightly more conservative figure.

29. Jones 2021, Figure 2, 15.

30. Jones 2021, Figure 2, 15. The claim that population growth also increases *per capita* incomes (rather than contributing to GDP just by increasing the number of workers) is precisely the essence of semiendogenous growth theory: more people find more ideas, which, because of their nonrival nature, make everyone more productive.

31. Geologists would say that we still *are* in an ice age—which they define as a period during which there are polar ice sheets and glaciers on Earth. More detail and references for the mutual isolation of the five regions at whatweowethefuture.com/notes.

32. Kremer 1993, 709. One caveat is that these regions started out with significant technological differences in 10,000 BC. For instance, there was agriculture in Mesopotamia but nowhere else (Stephens et al. 2019, Figure S2). Given the described outcomes in AD 1500, it does still seem correct that technological differences increased rather than decreased.

33. Sources differ on the exact numbers for 10,000 BC and AD 1, so I give only approximate figures here. For an overview of different estimates, see Our World in Data (2019a).

34. Jones 2001; Mokyr 2016.

35. In 2019, 3.1 percent of US GDP was spent on R&D (OECD 2021b). However, Jones and Summers (2020, 19) suggest that is likely a too-conservative accounting. In a survey they cite, firms report that only 55 percent of innovation costs were captured by R&D expenditures. In addition, things like venture capital investments in start-ups should arguably count as R&D investments, but this is captured by the official R&D figures only in part. Therefore, I adjust the 3 percent from the OECD upwards to account for some of these dynamics.

36. UN 2019b; Vollset et al. 2020, Figure 5, 1296.

37. In Figures 7.3 and 7.4, "live births per woman" more precisely refers to the total fertility rate (TFR). More at whatweowethefuture.com/notes.

38. World Bank (2021b, country-level data for 2019) and UN (2019b, average for high-income countries, 2015–2020).

39. *Economist* 2018; Vollset et al. 2020, Figure 8, 1299. Because of so-called population momentum, population levels can lag behind changes in the fertility rate. For example, if a population had been growing rapidly before the fertility rate fell below replacement level, the population can keep growing for a while as larger, later (middle-aged) cohorts replace smaller, earlier (older) cohorts. In the long run, though, if fertility rates are below replacement, the population will shrink.

40. https://population.un.org/wpp/Download/Standard/Fertility/. In 2020, China's fertility rate may have fallen to 1.3 (Marois et al. 2021, 1) and India's to 2.0—for the first time below the replacement rate (NFHS 2021, 3). It remains to be seen whether these declines are a temporary effect of the COVID-19 pandemic.

41. Vollset et al. 2020, 1290ff and Figure 3B, 1295. Significant population growth is projected for Australia as well, but this is an anomaly that's due to unusually high immigration. On the regional level, Central Asia is also projected to see sustained population growth this century, but in the long run the same remarks as for Africa apply.

42. Vollset et al. 2020, Figure 5, 1296; Bricker and Ibbitson 2019.

43. Vollset et al. 2020, Figure 3, 1295.

44. Vollset et al. 2020, 1285, 1290ff.

45. See whatweowethefuture.com/notes.

46. Walker 2020; Witte 2019; OECD 2020; Szikra 2014, 494–495.

47. World Bank 2021a.

48. See also Jones 2021, Section 6.2.

49. In Chapter 4, I gave an overview of multiple lines of evidence on the time until AGI (expert surveys: Grace et al. 2016; Zhang et al. 2021; comparisons with biological systems: Cotra 2020; reference-class forecasting: Davidson 2021a). There, I focused on the observation that they all agree that it is at least plausible that AGI will be developed soon—perhaps a 10 percent chance by 2036 and a 50 percent chance by 2050. However, this falls far short of establishing that we should expect AGI this century with very high confidence: Davidson's (2021a) reference class–based estimate is that "pr(AGI by 2100) ranges from 5% to 35%, with my central estimate around 20%"; Cotra (2020, Part 4, 17) concludes she could see herself "arriving at a view that assigns anywhere from ~60% to ~90% probability that TAI [a notion similar to AGI] is developed this century"; and disagreement among the experts surveyed by Grace et al. and Zhang et al. was so large that several thought it was less likely we'd see AGI within one hundred years, and even looking at the mean forecast rather than focussing on the pessimists among the respondents suggests a chance of at least 25 percent that AGI is more than a hundred years away. For qualitative experts' views on remaining challenges on the path to AGI, see Cremer (2021).

50. One of the authors of a study reporting the cloning of macaque monkeys, Mu-Ming Poo, said in 2018 that "technically, there is no barrier to human cloning" (quoted in Cyranoski 2018, 387).

51. Bouscasse et al. 2021.

52. I previously suggested that AGI provides a mechanism by which a lock-in of values could become permanent. But in this period of stagnation, we wouldn't have AGI yet—since if we had AGI, we wouldn't be stagnating. Without AGI, we should still expect cultural change over the course of many thousands of years. Over time, eventually, some culture that restarts growth will emerge.

53. See, for instance, Neilson's (2005) edited volume on "the Stark argument"—Rodney Stark's (1984, 18) contention that "the Mormons . . . will soon achieve a worldwide

following comparable to that of Islam, Buddhism, Christianity, Hinduism, and the other dominant world faiths." Note, however, that Stark's argument relies more on the Mormons' successful missionary efforts than their unusually high fertility: "One reason for Mormon growth is that their fertility is sufficiently high to offset both mortality and defection. But a more important reason is a rapid rate of conversion. Indeed, the majority of Mormons today were not born in the faith, but were converted to it" (Stark 1984, 22). Kaufmann (2010, 30) indicates this was still true more recently but also notes that "endogenous growth [i.e., from higher birth rates] is often more enduring" because "rapid conversion is often accompanied by rapid exit."

54. Perlich 2016.

55. Arenberg et al. 2021, 3–5.

56. Makdisi 1973, 155–168; Gibb 1982, 3–33; Bisin et al. 2019.

57. See whatweowethefuture.com/notes.

58. Indeed, getting back to sufficient population size to drive technological progress would take long enough that there would be no immediate reward of population growth from new technological innovation—population growth would only make a country richer in hundreds of years' time. So the population increase would have to happen for reasons other than purely economic incentives.

59. For the classic statement of this argument, see Bostrom (2003). See also Christiano (2013).

60. Friedman 2005.

61. Note that this consideration is also relevant for the discussion of the collapse of civilisation. If civilisation were to collapse, then, even if we were to recover eventually, the world would be guided by very different values than it is today.

62. Ord 2020, Table 6.1.

63. Note that the risk incurred during the period of stagnation would be purely additive. After we emerged from the stagnation, we'd still have to manage all the risk that we'd have incurred during the previous time if we had averted stagnation (e.g., risk from future technologies).

Chapter 8: Is It Good to Make Happy People?

1. Information for this section comes from personal acquaintances and Dancy (2020); Edmonds (2014); Srinivasan (2017); McMahan (2017; personal correspondence, October 12, 2021); MacFarquhar (2011). Parfit's wife, Janet Radcliffe-Richards, also an eminent moral philosopher, once commented that "Derek has no idea what it is for a building to exist without a manciple and domestic bursar" (quoted in Edmonds 2014).

2. Student Statistics 2021.

3. Colson 2016.

4. This stopped a few years ago (ASC 2021).

5. MacFarquhar 2011.

6. The Mohists, the Chinese school of thought I discussed in Chapter 4, argued that the good consisted of material prosperity, a large population, and social and political order. However, they did not discuss the intrinsic and instrumental benefits and costs of increasing population, and so they did not engage in population ethics in the sense I am interested in here (Fraser 2020; personal correspondence, October 11, 2021). More detail on the history of population ethics is available at whatweowethefuture.com/notes.

7. Parfit 1984, 453.

8. Parfit 2011, 620.

9. Narveson 1973, 80.

10. Broome 2004, Chapter 10. Krister Bykvist was my other supervisor.

11. Broome 2004, Preface. Confirmed in personal communication (November 25, 2021).

12. Huemer 2008, Section 4.

13. Caviola et al. 2022.

14. Parfit 1984, Chapter 16.

15. The situation is somewhat different for in vitro fertilisation (IVF) pregnancies. More at whatweowethefuture.com/notes.

16. "When people talk about traveling to the past, they worry about radically changing the present by doing something small, but barely anyone in the present really thinks that they can radically change the future by doing something small." I thank the Reddit forum r/Showerthoughts, with a hat tip to Brian Christian. (The quote is from user u/MegaGrimer, December 2, 2017; a very similar thought had been posted by u/kai1998 on November 5, 2016.)

17. Assuming that on average a person conceives one child in their lifetime, then a conception event occurs about once every twenty-nine thousand person-days.

18. Broome 2004, Chapter 10; Greaves 2017.

19. Roberts 2021.

20. Parfit 1984, 378–441.

21. Broome 1996, Section 4. For example: "If population growth and per capita GDP growth are completely independent, higher population growth rates would clearly lead to higher economic growth rates. It would still be true that, as noted by Piketty (2014), only the growth in per capita GDP would give rise to improvements in economic wellbeing" (Peterson 2017, 6). Ord (n.d.) discusses additional examples.

Caviola et al. (2022, 13, section 14.1.2.) asked participants which of a variety of different civilisations they thought were better. For example, they asked, "Civilization A contains 4,000 people at +60 happiness . . . Civilization B contains 6,000 people at +40 happiness . . . Which civilization is better?" On average, the respondents thought that Civilization A was better, even though both have the same total wellbeing—that is, the participants cared about the average wellbeing of the two civilisations.

22. An alternative version of the average view considers the average wellbeing of each generation at a time and regards a world as better if it has a higher sum of the average wellbeing of all generations. This, again, is a view that is sometimes assumed (implicitly or explicitly) by economists. However, it also has grave problems. For example, if we could choose, in the next generation, between a population of ten million people at wellbeing –100 or a population with those same ten million people and a further ten billion people at wellbeing –99.9, this view would recommend the latter because it would have the higher average well-being (Ord, n.d.).

23. Huemer 2008, Section 6.

24. Parfit 1984, Chapter 17.

25. Parfit 1986, 148.

26. Parfit 2016, 118.

27. Zuber et al. 2021.

28. Arrhenius 2000.

29. Blackorby and Donaldson 1984; Blackorby et al. 1997; Broome 2004.

30. There is an alternative version of the critical level view in which the addition of lives that are between zero and the critical level is not bad but neutral. This could be fleshed out in various ways, but one natural way is to say that, if two populations differ only insofar as one has an added life that is in between zero and the critical level, the two populations are incomparable in value (that is, neither is better than the other, nor are they equally good). One way to say this is to say that they are "on a par" (Chang 2002). Just to keep this discussion manageable, in this chapter I have put incomparability and parity to the side: I assume that the relation "is at least as good as" is complete.

31. Greaves 2017, Section 4.

32. MacAskill et al. 2020.

33. Greaves and Ord 2017. My colleagues Teruji Thomas and Christian Tarsney (2020) have shown that, in practice, other theories of population ethics converge in their implications to the critical level view.

34. Yglesias 2020, 52.

35. Wynes and Nicholas 2017.

36. Ord 2014.

37. I got the number of twenty billion galaxies from Ord (2020, 233). For an illuminating discussion of what we mean by the (currently) "affectable universe"—and how this notion differs from similar concepts, such as the observable universe, the eventually observable universe, and the ultimately observable universe—see Ord (2021).

38. For more on this idea, see Armstrong and Sandberg (2013).

Chapter 9: Will the Future Be Good or Bad?

1. If invertebrates are also sentient, then your sentient life would be enormously expanded: you would live for a hundred thousand trillion trillion years. Now, your time as a vertebrate would be a minuscule fraction of all of your experiences, and you would instead spend the vast majority of your time as nematodes, also known as roundworms, which live in the sea and on land.

2. The first vertebrate fossil is from the genus *Myllokunmingia*, around 520 million years ago, but there are several other candidate stem vertebrates (Shu et al. 1999; Donoghue and Purnell 2005, Box 2).

3. For estimates of the numbers in the "sentience as a single life" thought experiment, see whatweowethefuture.com/notes.

4. Schopenhauer 1974, 299.

5. Benatar 2006, 164. I'm grateful to Andreas Mogensen for pointing me to the quoted statements by Schopenhauer and Benatar, and more broadly for highly insightful conversations about the subject matter of this chapter.

6. Parfit 2011, 616–618.

7. WHO 2021a.

8. Our World in Data 2021a.

9. The median annual global income is $2,438 per year. In the UK, the median income for a full-time worker is £31,772 (Francis-Devine 2021).

10. Crisp 2021.

11. Diener et al. 2018a.

12. A fourth are surveys asking people about the balance of positive and negative emotions in their life. I have left this out because they seem particularly unhelpful because they don't weight by intensity of affect.

13. This is known as Cantril's self-anchoring striving scale.

14. Diener et al. 2018a; Diener et al. 2018b, 168. Their reported conclusions on positive affect were more upbeat, finding that "74% of respondents . . . felt more positive feelings . . . than negative feelings 'yesterday,' whereas only 18% . . . felt more negative feelings . . . than positive feelings 'yesterday.'" However, these results are particularly difficult to interpret: the measure of "more positive than negative feelings" was given by taking the average number of yes responses to two positive-affect questions (whether people smiled or laughed and whether they felt enjoyment much in the previous day) and subtracting the average number of yes responses to four negative-affect questions (whether people felt worry, sadness, depression, and anger much in the previous day). We can only say that the balance of affect was positive if we assume that the intensity of reported positive and negative affect was the same, on average. But this doesn't seem well motivated: for example, the intensity of positive affect required to smile or laugh once during a day seems much less than the intensity of negative affect required to say that one felt depression during a day.

15. Ng 2008.

16. This is known as the "reference group effect" (Credé et al. 2010).

17. Ponocny et al. 2016, Table 3. Note that this is different to "hedonic adaptation," which occurs when, after a chance in external life circumstances, someone returns to their previous, stable level of internal emotional state. For example, someone who permanently injured their leg in an accident might initially be quite unhappy, but over time they might hedonically adapt to their new condition and return to the level of happiness that they had previously had. Equally, someone who gets a promotion might initially be happier, but after a year or so their happiness would return to its prior state.

18. Ghana, Kenya: Redfern et al. 2019, 92f; UK: Peasgood et al. 2009, 7–11.

19. Helliwell et al. 2017, 14, Figure 2.1, shows that about 5 percent of the world population report a life satisfaction level of 0 or 1, and a further 5 percent a life satisfaction level of 2. More detail at whatweowethefuture.com/notes.

20. Ortiz-Ospina and Roser 2017.

21. Haybron 2008, 214–221.

22. Johansson et al. 2013.

23. Killingsworth et al. 2020.

24. One possible limitation of this study is that it might, in part, merely measure people's impatience to get to the next experience, rather than their judgment that an experience is not worth having at all. I might want to skip a car journey to a theme park, even if I am enjoying the car journey, because I would enjoy the theme park even more. Though the impulse to skip here is quite natural, it is also irrational. Whether I skip the car journey or not, I will still get to experience the theme park, so by skipping all I am doing is depriving myself of a positive experience—in effect, I am reducing my waking life expectancy for zero benefit. However, for a couple of reasons, it doesn't look like that is what is going on here.

First, the smaller study, which used the retrospective day reconstruction method, found that people skipped a similar amount of time as in the larger experience-sampling study. But impatience is not plausibly at play when we are retrospectively assessing which experiences we would prefer to have skipped. Second, if it is true that people want to skip to the next experience provided it is better, then the skipping method would not measure the absolute value of different experiences but rather their relative ranking for one person. Taken to its extreme, this claim would predict that the happy and the unhappy would skip the same number of experiences. However, the data shows the opposite. Skip percentage is highly correlated to how happy people are on average: the happier someone is, the less they want to skip. This suggests that skipping is not tracking impatience to get to the next relatively better experience; instead, it's tracking some judgment of whether an experience is worth having at all (Matt Killingsworth, personal communication, September 28, 2021).

25. And, based on personal correspondence with them (December 24, 2020; December 29, 2020; December 31, 2020; January 3, 2021; January 4, 2021), the authors think similarly.

26. Bertrand and Kamenica 2018.

27. Caviola et al. 2021.

28. Ortiz-Ospina and Roser 2017.

29. Easterlin 1974.

30. Easterlin and O'Connor 2020.

31. Stevenson and Wolfers 2008.

32. The vertical axis in Figure 9.1 refers to answers to the following question (English version): "Please imagine a ladder, with steps numbered from 0 at the bottom to 10 at the top. The top of the ladder represents the best possible life for you and the bottom of the

ladder represents the worst possible life for you. On which step of the ladder would you say you personally feel you stand at this time?" (Helliwell et al. 2021, 1)

33. Chan 2016.

34. Lustig's secret is reinvesting his winnings: "[Playing the lotto is] like any investment," Lustig said in one interview. "You have to invest money to get something out of it. Most people buy a one-dollar ticket and win ten dollars and they put the ten dollars in their pocket." Those people are playing the game wrong, he says. Instead, he advises, if you win ten dollars, then you should buy eleven dollars' worth of tickets because "if you lose, you only lost one dollar." It is unclear whether Lustig's net winnings are positive or not (Little 2010). Lustig has also released Lottery Maximiser software, which retails for ninety-seven dollars, and he ran an online course called Lottery Winner University.

35. Oswald and Winkelmann 2019. Earlier research found that winning the lottery had a small effect, but that research used a smaller sample than Oswald and Winkelmann.

36. For the definition and history of extreme poverty, see note 16 in Chapter 1.

37. Clark et al. 2016.

38. Stevenson and Wolfers 2008.

39. Roser and Nagdy 2014.

40. Dahlgreen 2016.

41. Mummert et al. 2011.

42. This argument became especially prominent in the 1970s, with Marshall Sahlins's (1972) notion of an "original affluent society." Sahlins's argument is controversial (see, e.g., Kaplan 2000). For a more pessimistic take on preagricultural quality of life, see Karnofsky (2021e).

43. Kelly 2013, 12–14.

44. Kelly 2013, 243ff. See also Marlowe 2010, 43ff.

45. Marlowe 2010, 67f.

46. National Geographic Society 2019. References on diet at whatweowethefuture.com /notes.

47. Frackowiak et al. 2020, Table 4. See also Biswas-Diener et al. 2005. Williams and Cooper (2017) find rural Himba participants who practice a traditional seminomadic pastoralist lifestyle have higher scores on the Satisfaction with Life Scale than a matched sample of UK participants.

48. Kelly 2013, Chapter 10.

49. Kelly 2013, Table 7.8.

50. For example, Turnbull 2015; Everett 2008; Marlowe 2010; Lee 1979; Rival 2016; Suzman 2017.

51. Volk and Atkinson 2013, Table 1; Our World in Data 2018b.

52. UK Office for National Statistics 2019.

53. For an overview, see Kelly (2013, Chapter 7). For arguments in favour, see Pinker (2012). For arguments against, see Lee (2018); Fry (2013).

54. Our World in Data 2020d.

55. Christensen et al. 2018.

56. These data are from FAOSTAT data set maintained by the Food and Agriculture Organization of the United Nations (FAO), summarised by Šimčikas (n.d.).

57. Many more fish die before being slaughtered.

58. See, e.g., Humane Society of the United States 2009, 2013.

59. Based on official records of federally inspected slaughter plants in the United States, 440,000 chickens were scalded alive in that country in 2019. Since the United States accounts for one-seventh of global meat consumption and has above-average welfare standards, millions of chickens probably die in this way each year worldwide (National Agricultural Statistics Service 2021).

60. Other farmed animals suffer even worse fates, such as ducks and geese raised to make foie gras: "Ducks and geese are force-fed via a long tube inserted down their esophagi with an unnatural quantity of food pumped directly into their stomachs. . . . Birds force-fed for foie gras may suffer from a number of significant welfare problems, including frustration of natural behavior, injury, liver disease, lameness, and diseases of the respiratory and digestive tracts, and higher rates of mortality compared to non force-fed ducks" (Humane Society of the United States 2009, 2).

61. Compassion in World Farming 2021.

62. Animal Charity Evaluators 2020, Appendix, Table 4.

63. Compassion in World Farming 2009, 12.

64. Mood and Brooke 2012, 22f; Poli et al. 2005, 37.

65. Compassion in World Farming 2021.

66. Among the rare exceptions is Bailey Norwood, who in *Compassion by the Pound* claims that most broiler chickens have positive wellbeing. His coauthor, Jayson Lusk, has a different view on this question (Norwood and Lusk 2011, Chapter 8).

67. For a review of this question, see Schukraft (2020).

68. Bar-On et al. 2018, 6507, Figure 1.

69. Polilov 2008, 30; Menzel and Giurfa 2001, 62; Olkowicz et al. 2016, Table S1; Azevedo et al. 2009.

70. Note that this is only true if we exclude invertebrates. If we included them and used simple neuron count, then we would conclude that our attention should be entirely focused on nematodes.

71. Bar-On et al. 2018, Supplementary Information 36f.

72. Bar-On et al. 2018, Figure 1.

73. Bar-On et al. 2018, Supplementary Information 34–36.

74. Triki et al. 2020, 3, assuming half the brain cells are neurons.

75. Houde 2002, 68f. The common carp can live up to thirty-eight years and the wels catfish up to eighty years (Froese and Pauly 2021ab).

76. Houde 2002, Section 3.3.

77. Some people even argue that many animals in captivity have better lives than wild animals. Various studies have shown that wild animals have higher levels of the stress hormone cortisol than domesticated animals (Wilcox 2011; Davies 2021, 307–313).

78. This is also the finding of a recent paper on wild animal welfare (Groff and Ng 2019, 40). In the previous chapter I suggested that, under moral uncertainty, we should follow something close to a critical level view of population ethics. If this is correct, then we should regard the existence of most wild animals as a bad thing. Even if those animals have positive wellbeing lives, it seems very unlikely that they have sufficiently good lives to be above the critical level for wellbeing.

79. Bessei 2006, 10; Berg et al. 2000, 36; Knowles et al. 2008, Table 1.

80. Bar-On et al. 2018, 6508.

81. Christensen et al. 2014; Bar-On et al. 2018.

82. Ritchie and Roser 2021a; Dirzo et al. 2014, 401–406; Tomasik 2017, 2018.

83. Hurka 2021; Brennan and Lo 2021.

84. Ritchie and Roser 2021c.

85. Ritchie and Roser 2021a; McCallum 2015, 2512.

86. Roser 2013a.

87. Roser 2013d.

88. Russell 2010, 1.

89. Quoted in Yarmolinsky 1957, 158.

90. This paragraph borrows from an excellent blog post by Althaus and Baumann (2020).

91. Chang and Halliday 2006, Chapters 8, 23, 48.
92. Glad 2002, 14.
93. I'm grateful to Carl Shulman for making this point to me.

Chapter 10: What to Do

1. This section is based on Núñez and Sweetser (2006). Aymara is the best-studied exception to the rule, but there may be others. According to one study, in Vietnamese, time can approach from behind and "continue forward" into the past (Sullivan and Bui 2016). The Yupno represent time as running uphill and downhill, and the Pormpuraawan people conceptualise it as running east to west (Núñez et al. 2012; Boroditsky and Gaby 2010).

2. *Encyclopedia Britannica* 2016. There are more than 1 million Aymara in Bolivia (Instituto Nacional de Estadística 2015, Cuadro 7), half a million in Peru (Instituto Nacional de Estadistica e Informatica 2018, Cuadro 2.69), 150,000 in Chile (Instituto Nacional de Estadisticas 2018, 16), and 20,000 in Argentina (Instituto Nacional de Estadistica y Censos 2012, 281).

3. Indeed, it's plausible that the Aymara language has this idiosyncratic conceptual metaphor because in general it incorporates a strong grammatical distinction, marked with verbal inflection or syntax, between knowledge gained via direct perception and knowledge gained secondhand. It's almost impossible to assert something in Aymara without indicating what your source is.

4. Clarke et al. 2021ab. The precise wording was, "Conditional on an existential catastrophe due to AI having occurred, please estimate the probability that this scenario occurred," for each of the six scenarios mentioned.

5. Clarke et al. 2021a.

6. Muehlhauser 2021.

7. Rumsfeld 2002.

8. CNN 2003.

9. Dartnell 2015a, 53f. The organisation ALLFED (https://allfed.info) is working on developing food production that doesn't require sunlight.

10. A 2021 survey of AI researchers and leading institutes reached out to 135 researchers, so 120 is a plausible lower bound (Clarke et al. 2021ab). The main funder in the space is Open Philanthropy, which donates tens of millions in this area each year (see their grants database in Open Philanthropy 2021).

11. For a longer list of topics, see GPI's research agenda at https://globalprioritiesinstitute.org/research-agenda/.

12. CFCs were the main contributor to the ozone problem, but other ozone-depleting substances were also important (Ritchie and Roser 2018a).

13. DuPont alone had about a quarter of the global CFC market, and the global market was dominated by only five companies. The market was only worth $600 million. For comparison, the market for fossil fuels is worth trillions (Falkner 2009, 52). The CFC substitutes only increased short-term costs by a factor of two to three (US National Academy of Sciences 1992).

14. Molina and Rowland first published their paper on the connection between CFCs and the ozone layer in 1974. They later won the Nobel Prize in Chemistry. The Montreal Protocol came into force in 1989 (Ritchie and Roser 2018a). Today, emissions of CFCs and other ozone-depleting substances have fallen to close to zero. The ozone hole stabilised in the 1990s and started to shrink around 2005 (Ritchie and Roser 2018a).

15. For an outline of the political economy challenges of climate change, see Cullenward and Victor (2020).

16. In 2019, global philanthropic spending on climate change was $5 billion to $9 billion (Roeyer et al. 2020). For the amount spent by governments and companies, see UN 2021a.

Around a third of young people in the United States say that addressing climate change is their top personal concern (Tyson et al. 2021).

17. Wynes and Nicholas 2017, Supplementary Materials 4, Figure 17.

18. Ritchie and Roser 2018b.

19. Wynes and Nicholas 2017, Supplementary Materials 4.

20. Note that this is consumption-based emissions per capita. This accounts for the carbon dioxide released when we purchase products that are made overseas using fossil energy. For the UK, that figure is 7.7 tonnes of carbon dioxide per capita annually (Ritchie 2019).

21. If those numbers sound unbelievably low given how expensive directly reducing emissions often is, consider that they are the result of leveraging various impact multipliers, namely: using advocacy to improve governments' resource allocation, doing so for climate technologies that are otherwise neglected, and using innovation to discover solutions that can then be implemented globally. In 2018, Founders Pledge estimated that the Clean Air Task Force has historically averted a tonne of carbon dioxide for around one dollar per tonne and predicted that the cost-effectiveness of their future projects would be higher (Halstead 2018a, Section 3.2).

22. Van Beurden 2019.

23. Wiblin 2020; Edlin et al. 2007.

24. Schein et al. 2020; Green and McClellan 2020.

25. Quoidbach et al. 2013, Supplementary Materials 6; Orr 2015.

26. Wiblin and Harris 2019.

27. Todd 2021a, based on Daniel and Todd 2021.

28. Todd 2021b, n1.

29. See also Karnofsky 2021a.

30. BBC 2021.

31. Yan 2021.

32. MacAskill et al. 2020.

33. Gerbner 2007.

34. Ford 2010.

35. *Encyclopedia Britannica* 2020f. The Representation of the People Act gave British women the vote in 1918, but while all men over the age of twenty-one could vote, women had to be over thirty and meet a property qualification. The 1928 Equal Franchise Act finally gave men and women equal voting rights (UK Parliament 2021a).

36. Note that this is a fraction of total deployment of solar capacity, not actual solar generation (Sivaram 2018, 36).

37. Our World in Data 2019h.

38. The effect of the environmental movement in Germany has not been wholly positive, however. As well as increasing support for solar power, the Greens also advocated for nuclear power to be phased out entirely, which did great damage to the climate because the outgoing nuclear power was largely replaced by coal. On some estimates, because of the extra air pollution, an additional 1,100 people died every year as a result of this policy decision (Jarvis et al. 2019).

39. UN 2021b. Toby Ord's (2020) *The Precipice* is among the references.

40. Demand for *useful energy*, which Way et al. (2021, 9) define as "the portion of final energy used to perform energy services, such as heat, light and kinetic energy," has historically grown 2 percent per year. Because fossil fuels waste a lot of energy compared to renewable electricity, low-carbon electricity supply may not need to grow at 2 percent per year to meet rising useful-energy demand.

41. Pirkei Avot 1:14 as quoted in Carmi (n.d.).

42. Roser 2013d.

Appendices

1. Greaves and MacAskill 2021.

2. Mogensen 2020. On very small probabilities, see Beckstead and Thomas (2021); Tarsney (2020a); Wilkinson (2020, forthcoming); and Bostrom (2009); and for an accessible discussion, see Kokotajlo (2018). On acting in the face of ambiguous evidence, see Lenman (2000); Greaves (2016); Mogensen (2021); Tarsney (2020b [2019]); and Cowen (2006).

3. Ord 2020. See also Bostrom 2002, 2013.

4. I'll later talk about p and q as possible worlds, but really all that's required is that $V_s(p)$, $V_s(q)$, $T_s(p)$, and $T_s(q)$ are well defined. That is, p and q could also be propositions that specify (at least) for how long the world would be in state s and how much value this would contribute. (Such propositions could in turn be cashed out as sets of possible worlds in which they are true, though this is not required to use the SPC framework.)

5. I use this example to illustrate, although the claim that QWERTY keyboards are an example of bad lock-in seems spurious. It's often claimed that the QWERTY layout was designed to slow down users of typewriters in order to prevent jams, but this is an urban legend. And evidence for Dvorak's superiority is scant; rather, Dvorak's reputation seems to be largely the product of advertising and biased studies run by August Dvorak himself (Liebowitz and Margolis 1990).

6. Here I assume that the value contributed by Dvorak being the standard in period 4 is the same in worlds X and O (see Table). The requirement that the value contributed by the state s under consideration only depends on how long the world is in that state should perhaps be added to the definition of the SPC framework, since otherwise it doesn't make much sense to use $T_s(p) - T_s(p)$ in the definitions of significance and contingency.

7. There are two possible sources of uncertainty. First, we might be uncertain about the effect p of the action under consideration. Second, we might be uncertain about the status quo q.

8. Open Philanthropy, n.d.

9. This is a variation of a formalization by Owen Cotton-Barratt (2016) and was suggested to me by Teruji Thomas. The two formalizations differ substantively in how they cash out tractability and neglectedness. More at whatweowethefuture.com/notes.

10. Beckstead and Thomas 2021.

11. On moral uncertainty, see MacAskill et al. 2020.

12. See the discussion in Chapter 1.

13. Back-of-the-envelope calculations suggest that some actions to avert permanent catastrophes might be unusually cost-effective even when compared to many activities aimed at improving the quality of life of people today, such as health-care spending in rich countries (Lewis 2018). See also Wiblin and Harris (2021, October 5).

Index

Afterwards

A story of a good future.

For Holly

Credit: Matt Crockett

William MacAskill is an associate professor in philosophy and senior research fellow at the Global Priorities Institute, University of Oxford. At the time of his appointment, he was the youngest associate professor of philosophy in the world. He has focused his research on moral uncertainty, effective altruism, and future generations. A TED speaker and past *Forbes* 30 Under 30 social entrepreneur, he also cofounded the nonprofits Giving What We Can, the Centre for Effective Altruism, and Y Combinator-backed 80,000 Hours, which together have moved over £200 million to effective charities. He is the author of *Doing Good Better* and lives in Oxford.